T0338264

Guidelines for

Process Safety Knowledge Management

Center for Chemical Process Safety

American Institute of Chemical Engineers

New York, NY

WILEY

For general information on our other products and services or for technical support, please contact our Customer Care Department within the United States at (800) 762-2974, outside the United States at (317) 572-3993 or fax (317) 572-4002.

Wiley also publishes its books in a variety of electronic formats. Some content that appears in print may not be available in electronic formats. For more information about Wiley products, visit our web site at www.wiley.com.

Library of Congress Cataloging-in-Publication Data applied for:
Hardbook ISBN: 9781394187713

Cover Images: Silhouette, oil refinery © manyx31/iStock. com; Stainless steel © Creativ Studio Heinemann/Getty Images; Dow Chemical Operations, Stade, Germany/Courtesy of The Dow Chemical Company

SKY10068380_030124

Guidelines for

Process Safety Knowledge Management

This book is one in a series of process safety guidelines and concept books published by the Center for Chemical Process Safety (CCPS). Please refer to wiley.com/go/ccps for a full list of titles in this series.

Table of Contents

List of Figures

List of Tables

Acronyms and Abbreviations

AIChE	American Institute of Chemical Engineers
API	American Petroleum Institute
CCPS	Center for Chemical Process Safety
CRW	Chemical Reactivity Worksheet
DIKW	Data-Information-Knowledge-Wisdom
HAZID	Hazard Identification
HAZOP	Hazards and Operability Study
KM	Knowledge Management
KPI	Key Performance Indicators
MOC	Management of Change
MOOC	Management of Organizational Change
PHA	Process Hazards Analysis
PSI	Process Safety Information
PSK	Process Safety Knowledge
PSKM	Process Safety Knowledge Management
RACI	Responsible, Accountable, Consulted, and Informed chart
RAGAGEP	Recognized and Generally Accepted Good Engineering Practices
RBPS	Risk Based Process Safety (CCPS)
SDS	Safety Data Sheet

Glossary

This Glossary contains Process Safety terms significant to this CCPS publication, which are current at the time of publication. For other CCPS Process Safety terms and updates to these terms, please refer to the CCPS Process Safety Glossary [1].

Term	Definition
Accident precursors [2]	Events that must occur for an accident to happen in each scenario but have not resulted in an accident so far
Cause (Incident)	An event, situation, or condition which results, or could result (Potential Cause), directly or indirectly in an accident or incident [1].
Chief Knowledge Officer	Person accountable for the overall PSKM strategy, planning and implementation (Highest position within PSKM)
Contributing Cause	Factors that facilitate the occurrence of an incident such as physical conditions and management practices (also known as contributory factors) [1].
Key Performance Indicators (KPI)	A quantifiable way to monitor the health of the overall PSKM System and proactively identify potential issues early to be corrected or improved. KPIs tell an organization how effective their PSKM is at supporting their RBPS program.
Logic Model	A logic model is a tool that can be used to develop and implement the PSKM System [3]. Logic models are graphic illustrations of the PSKM Implementation Plan and show the relationship between the planned work and anticipated results.
Management of Organizational Change (MOOC)	Framework for managing the effect of new business processes, changes in organizational structure or cultural changes within an organization

Term	Definition
Process Safety Knowledge (PSK)	Knowledge is related to information, which is often associated with policies, and other rule-based facts. It includes work activities to gather, organize, maintain, and provide information to other process safety elements. Process Safety Knowledge primarily consists of written documents such as hazard information, process technology information, and equipment-specific information.
Process Safety Knowledge Management (PSKM)	System for capturing, organizing, maintaining, and providing the right Process Safety Knowledge to the right people at the right time to improve process safety in an organization
Process Safety Knowledge Management Focus Chart	A chart divided into three columns that depict causes and other factors related to the incident, and four rows that show elements of the PSKM System (i.e., Capture, Organize, Maintain, and Provide).
Process Safety Knowledge Management System	A tool that makes necessary Process Safety Knowledge available to everyone who needs it, when they need it, and at the right level of detail
Proximate Cause	The cause factor which directly produces the effect without the intervention of any other cause. The cause nearest to the effect in time and space [1].
PSKM Audit	A PSKM audit expands on a regulatory audit such that it covers not only availability of documents but their content, accuracy, system/process to create/update, and how the information is shared and utilized. A PSKM audit benefits an organization by identifying gaps in the system and improvement opportunities.
PSKM Champions	A PSKM Champion will promote PSKM in the workplace and facilitate Communities of Practice.

Term	Definition
PSKM Editors	A PSKM Editor is someone who knows where PSKM is located and manages format and language of knowledge so users can easily use it.
PSKM Navigators	Navigators connect people who need knowledge with systems and people who have knowledge.
PSKM Project Manager	An executive who manages the implementation of the PSKM initiatives
PSKM Stewards	A steward is responsible for ensuring PSKM updates are made following Management of Change and track changes for follow-up and validation.
Root Cause	A fundamental, underlying, system-related reason why an incident occurred that identifies a correctable failure(s) in management systems. There is typically more than one root cause for every process safety incident [1].

Acknowledgments

The American Institute of Chemical Engineers (AIChE) and the Center for Chemical Process Safety (CCPS) express their appreciation and gratitude to all members of the *Guidelines for Process Safety Knowledge Management* Subcommittee for their generous efforts in the development and preparation of this important guideline. CCPS also wishes to thank the subcommittee members' respective companies for supporting their involvement during the different phases in this project.

Subcommittee Members:

Michelle Brown, Chair	FMC
Denise Albrecht, Co-Chair	3M
Jennifer Brittain	AdvanSix
Brian Farrell	CCPS Consultant
Linus Hakkimattar	ReVizions
Mark Hall	Mallinckrodt Pharmaceuticals
Dan Hannewald	BASF
Rainer Hoff	Gateway Group
Allison Knight	3M
Jennifer Mize	Eastman chemical
Steve Murphy	Syngenta
Mohammad Nashwan	Saudi Aramco
Ravi Ramasamy	Nghi Son Refinery & Petrochemical LLC
Jeffery Todd	Holly Frontier
Florine Vincik	BASF
Jerry Yuan	IRC Risk
Hafeez Ahmad Zeeshan	Tronox Management Pty Ltd

The book committee wishes to express their sincere appreciation to PSRG (Robert J. Weber, Tekin Kunt, Madonna Breen, Michael Munsil, Ester Zelaya, Aaran Green, Ngoc "Annie" Nguyen, Carolina Del Din, Jimmy D Trinh, Russ Kawai, Ryan Terry, and Sonny Sachdeva) for their contributions in preparing the guideline's manuscripts.

Before publication, all CCPS guidelines are subjected to a peer review process. CCPS gratefully acknowledges the thoughtful comments and suggestions of peer reviewers. Their work enhanced the accuracy and clarity of this guideline.

Although the peer reviewers provided comments and suggestions, they were not asked to endorse this guideline and did not review the final manuscript before its release.

Peer Reviewers:

Jack Chosnek	Knowledge One
Raj Dahiya	AIG
Emmanuelle Hagey	NOVA
Trish Kerin	IChemE
Joompote Ketkeaw	SCG Chemicals
Shannon Ross	Chevron
Juliana Schmitz	Linde
Herve Vaudrey	Dekra

Dedication

Kenneth E. Tague, CCPSC, CSP

Ken Tague is a Rose-Hulman Institute of Technology graduate with a career spanning over 38 years in chemical operations. His many roles have included Production Manager and Plant Manager. Before retirement, he was the CCPS Technical Steering Committee (TSC) representative for Archer Daniels Midland Company (ADM) and was on the CCPS Planning Board. His experience and presentation skills have made him a sought-after instructor for CCPS's flagship course, *Foundations of Risk Based Process Safety*. He has contributed to the AIChE SAChE and RAPID education programs by developing e-learning courses related to process safety.

Ken also served on CCPS book committees, contributing to the development of two CCPS books: *Dealing with Aging Process Facilities and Infrastructure* and *Recognizing and Responding to Normalization of Deviance*. Based on his hands-on experience, he also significantly contributed to the web-based training on *Process Safety for Maintenance Workers and Operators*.

He is a strong proponent of process safety, having shared his commitment to Process Safety at the 2018 Global Congress on Process Safety in the session "When PSM Hit Home." Preventable incidents continue to stir his passion for sharing his experiences to strengthen the expertise of engineers new to and within the Process Safety field.

Ken is a CCPS Certified Process Safety Professional (CCPSC), a Certified Safety Professional (CSP) in the Safety, Health, and Environmental (SH&E) field, and was an active member of the CCPS Pharma, Food, and Fine Chemicals (PFFC) Committee before he retired. In addition, he holds Patent 9,481,609 as a co-inventor of the process to make *Heteromorphic Lysine Feed Granules*.

CCPS is delighted to dedicate this book to Ken in recognition for his past, present, and continuing support of CCPS and the global Process Safety community.

Louisa A. Nara, CCPSC
Global Technical Director, CCPS

Anil Gokhale, Ph.D.
Chief Operating Officer, CCPS

Preface

The Center for Chemical Process Safety (CCPS) has been the world leader in developing and disseminating information on process safety management and technology since 1985. The CCPS, an industry technology alliance of the American Institute of Chemical Engineers (AIChE), has published over 100 books in its process safety guidelines and process safety concepts series, and over a hundred courses, including 33 training modules through its Safety in Chemical Engineering Education (SAChE) series. CCPS is supported by the contributions and voluntary participation of more than 250 companies globally.

This book contains guidelines for companies to improve their process safety performance through the implementation of a Process Safety Knowledge Management (PSKM) system. The characteristics of a PSKM system are defined and guidelines are shared on how to set up a PSKM system to improve overall Process Safety performance. The underlying factors for success are presented which include leadership, employee involvement, and organizational culture with case studies used to illustrate key points and learnings. New perspectives on PSKM are included along with strategies to overcome difficulties in transitioning from a process safety culture based on data and information to a culture based on knowledge and wisdom. Case studies with PSKM-related lessons learned demonstrate the principles and practices described in the book.

1 Introduction

"A society grows great when old people plant trees whose shade they know they shall never sit in." Greek proverb

1.1 What Is Process Safety Knowledge Management (PSKM)?

This chapter introduces the key definitions for Process Safety Knowledge and Knowledge Management. Process Safety Knowledge Management (PSKM) is a subset of Knowledge Management focusing on building, disseminating, and sustaining Process Safety Knowledge (PSK) in an organization.

Knowledge Management has been defined by many authors over the years as collected and published by Girard and Girard [4]. One of the classic and most cited definitions of Knowledge Management is by O'Dell and Grayson [5]:

"Knowledge Management is a conscious strategy of getting the right knowledge to the right people at the right time and helping people share and put information into action in ways that strive to improve organizational performance."

CCPS defines Process Safety Knowledge (PSK) as follows [1]:

"Knowledge related to information, which is often associated with policies, and other rule-based facts. It includes work activities to gather, organize, maintain, and provide information to other process safety elements. Process Safety Knowledge primarily consists of written documents such as hazard information, process technology information and equipment-specific information."

Hence, Process Safety Knowledge Management (PSKM) is defined as:

"A system for capturing, organizing, maintaining, and providing the right Process Safety Knowledge to the right people at the right time to improve process safety in an organization."

PSKM includes methodologies, tools, processes, organizational structures, and human capital management strategies used to convert data to information, information to knowledge and knowledge to wisdom.

Process Safety Knowledge Management (PSKM) systems cover the entire life cycle of Process Safety Knowledge including development, implementation, and maintenance. The knowledge management system must ensure Process Safety Knowledge is <u>easily accessible and understandable to the people who need it,</u>

and that the knowledge shared is consistent, current, and accurate. We note that there is a special case with regards to Contractors and other outside entities which could impact this stated goal. A brief discussion is included in Section 4.5.

1.2   Purpose and Scope of this Book

This book is intended to be a resource for sharing industry-leading best practices on PSKM and for providing a blueprint for developing an effective PSKM program for companies. This book is divided into three sections:

1. Business case for an effective PSKM program and its relationship to PSM elements (Chapters 2 and 3)
2. Setting up a successful PSKM system and sustaining it (Chapters 4 and 5)
3. Sharing case studies illustrating the importance of an adequate and effective PSKM system (Chapter 6)

The principles of PSKM are transferable across industries. Examples contained within this book will provide guidance on how the knowledge obtained from past incidents, and current best practices from industry leaders, can be applied to many different organizations.

1.3   Historical Development of PSKM

Historically, the terms Process Safety Information (PSI) and Process Safety Knowledge (PSK) have been used interchangeably [6], [7]. As companies' maturity level in Process Safety Management (PSM) improves, there is a continuing focus in the industry to transition from information to knowledge.

Before the 1990's, the PSK resided in the organization as a core competency of chemical or process engineers. Analysis of serious process safety events such as the methyl isocyanate release at Bhopal, India in 1984 and explosions at a chemical complex at Pasadena, Texas in 1989, showed that while PSI resided within an organization, it did not consistently turn into knowledge at the operational level. Hence, the right knowledge was not available to the right people.

With the establishment of the US PSM Standard by Occupational Safety and Health Administration (OSHA) in 1992, the importance of PSI and informing all affected employees of PSI became a key requirement in US facilities. Regulation, however, left the importance of Process Safety Information at the information level without extending it to knowledge.

Risk Based Process Safety incorporates four pillars: Commitment to Process Safety, Understand Hazards and Risk, Manage Risks and Learn from Experience. The Center for Chemical Process Safety (CCPS) thoroughly describes these pillars and 20 associated elements in its publication, Guidelines for Risk Based Process Safety (RBPS) [7]. One of the elements under the second RBPS pillar (Understand Hazards and Risk) is Process Knowledge Management. Process Safety Knowledge Management (PSKM) is different from the Process Knowledge Management (PKM) element. It directly impacts the four pillars of RBPS while operating within and beyond the limits of the individual elements of each pillar. In fact, PSKM is an expansion of PKM whereby it is actively used and maintained. PSKM is not just a catalog of information, but instead knowledge that is used to safely manage a process.

The Guidelines for Risk Based Process Safety (RBPS) book as published by CCPS in 2007 [7] highlighted the need for PSK and its inclusion in a wider management system. This work linked the PSK to the risk assessment and other risk management tools.

Initially, PSKM was based on word of mouth such as limited passing of anecdotal stories handed down from one to another almost like apprentices in trades. With the extension of PSI to PSK, there is an intentional and systematic processing of facts, records, and information management. It is important to understand the reasoning behind the action and not simply to replicate it.

1.4   Knowledge Pyramid

PSKM can be approached from the continuous improvement perspective of the Plan-Do-Check-Act (PDCA) cycle associated with other management systems. Using the PDCA cycle within the context of PSKM facilitates movement from simply providing information to people who operate and maintain processes to transferring knowledge and developing wisdom in these very same people.

The transition from data to wisdom can be represented by the Data-Information-Knowledge-Wisdom (DIKW) pyramid [8] (Figure 1-1).

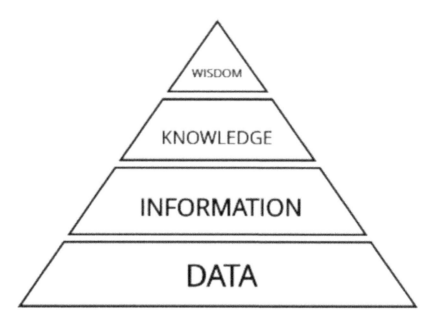

Figure 1-1 Data-Information-Knowledge-Wisdom (DIKW) pyramid

Data: There are millions of data items in a process facility. Data is a collection of *name-value* pairs. An example of a *name-value* pair is the "Normal Operating Pressure" as the name and "100 psig" as the *value*. When these name-value pairs are collected over time, they form data. For example, the Unit is Isomerization, Time Step is 5, and the Normal Operating Pressure is 100 psig. Data can usually be represented in a tabular format, either on paper or electronically.

Information: This is a grouping of data items, describing their interrelationships, along with suitable narratives. Examples of *information* include drawings (e.g., Process Flow Diagrams - PFDs, Piping, and Instrumentation Diagram - P&IDs), procedures (e.g., normal shutdown procedure, start-up procedure), equipment files (e.g., pump curves, tank design specifications), and inspection reports (e.g., wall thickness measurements, vibration tests, etc.). Information can be presented in tabular, graphical, or textual formats, either on paper or electronically.

Knowledge: When information is assimilated by a person it becomes *knowledge*. Reading and viewing information is a common way for people to acquire knowledge. Training is an organized approach to assure that information has been transformed into knowledge in specified individuals. Knowledge can be further categorized in two dimensions: explicit and tacit [9]. *Explicit knowledge* consists of facts, rules, relationships, and policies that can be faithfully codified in paper or electronic form and shared without need for discussion. *Tacit knowledge* represents knowledge based on the experience of individuals [10]. Tacit knowledge is knowledge housed in the human brain, such as expertise, understanding, or professional insight formed because of experience. Tacit knowledge is highly personal, context-specific, and therefore hard to formalize and communicate. Tacit knowledge in an organization is human capital, whereas explicit knowledge is structural capital.

Wisdom: The experience gained by applying knowledge in practice provides a deeper understanding of the knowledge, leading to competency and enhanced expertise and, when providing exceptional value, is normally termed "*wisdom*." Wisdom relates to the ability to effectively choose and apply appropriate knowledge in each situation. Wisdom is an action-oriented concept. Knowledge is required but insufficient for wise action. Organizational wisdom is the collection, transference, and integration of individuals' wisdom and the use of institutional and social processes (e.g., structure, culture, leadership), in understanding how a company makes best use of its knowledge [11].

Organizations that operate at the bottom of the pyramid tend to be very reactive. As they move up the pyramid, from information-based to knowledge-based, the organization becomes more independent and more proactive since it has developed personal and/or group knowledge it can use to operate more efficiently. The final move to organizational wisdom transforms the organizational culture to interdependence where understanding of process operations is deep and more effective.

There are multiple inputs and outputs for each of the DIKW pyramid sections. As shown in Table 1-1, as one moves up the pyramid the outputs of the previous stage become the inputs of the next stage – representing the development occurring within the organization as people transition from phase to phase.

Table 1-1 Example Inputs and Outputs of the DIKW Pyramid

	INPUTS	OUTPUTS
Data	Safety Data Sheets Test reports Equipment lists	P&ID Flow diagrams Process Safety Information Manuals Hazard Identification and Risk Analysis (HIRA)
Information	P&ID Flow diagrams PSI Manuals HIRA	Risk assessments Controls list Operating procedures Training Emergency plans
Knowledge	Risk assessments Controls list Operating procedures Training Emergency plans	Verification of understanding Validation of competency Revalidation plans Shared learnings External learnings
Wisdom	Verification of understanding Validation of competency Revalidation plans Shared learnings External learnings	Broader learnings Subject matter experts Knowledge sharing Knowledge retention Right decisions

1.5   Audience

This book is intended primarily for Engineering Managers and Process Safety Managers of all organizations who realize that a structured approach to Process Safety Knowledge Management can add value and improve safety. The intended audience also includes Process Safety Specialists, Plant Managers, Maintenance Managers, Operations Managers, and anyone else in the organization who has a role in improving process safety performance through a structured implementation of Process Safety Knowledge Management.

1.6   Elements Not Covered in This Book

The following elements are outside the scope of this book:

- Complete listing of all items of PSI to collect for a project or a facility
- Complete listing of all items of Process Knowledge to collect for a project or a facility
- Application of artificial intelligence and machine learning to development of PSK without human interaction
- Information security including cybersecurity
- Business knowledge management

This book relates directly to the Process Safety Knowledge Management (PSKM) instead of covering a more general topic of "Knowledge Management for any Chemical Business". As such, while business topics such as succession planning are touched upon in Chapters 4 and 5, this book is aimed specifically at understanding, implementing, and maintaining a PSKM System.

1.7   Content and Organization of this Book

The summaries provide an overview of the content and organization of this book, chapter-by-chapter, to enable the reader to quickly locate a particular area of interest.

This book is divided into three sections. The first section (Chapters 2 and 3) introduces the business case for Process Safety Knowledge Management along with its interaction with multiple elements of Risk Based Process Safety. The second section (Chapters 4 and 5) addresses how to set up and sustain a successful Process Safety Knowledge Management system. The last section (Chapter 6) provides case studies and lessons learned from various industries.

Chapter 2 – Business Case for Process Safety Knowledge Management: this chapter discusses the benefits of having an effective Process Safety Knowledge Management system in an organization.

Chapter 3 – Process Safety Knowledge Management and Risk Based Process Safety: this chapter further describes Process Safety Knowledge Management and its interaction with relevant elements of Risk Based Process Safety such as Hazard Identification and Risk Analysis, Management of Change, Incident Investigation, and others.

Chapter 4 – Developing and Implementing Process Safety Knowledge Management: this chapter discusses subjects about developing and implementing a Process Safety Knowledge Management system relevant to facility design, operations, and maintenance.

Chapter 5 – Maintaining and Improving Process Safety Knowledge Management: this chapter builds on Chapter 4 content to provide a summary of the applicable tools currently available for Process Safety Knowledge Management. In this chapter, various tools are introduced to illustrate the principles described in Chapter 4. This chapter also highlights available metrics related to managing Process Safety Knowledge that can be used for improvement, as well as their ongoing review for a sustainable PSKM system.

Chapter 6 – Case Studies and Lessons Learned: this chapter includes Case Studies from around the world and their relevance to Process Safety Knowledge Management. The Case Studies share general as well as specific information regarding Process Safety Knowledge, related to the events that have occurred both within and outside the chemical process industry. The Case Studies describe how these events were handled from the perspective of management of Process Safety Knowledge. Effective and inadequate examples of PSKM are included in concise summaries.

2 The Business Case for Process Safety Knowledge Management

"Process safety must be lived. It can't be only a binder on a shelf." [12]
Kimberly McHugh, VP Drilling & Completions, Chevron

Process Safety Knowledge Management (PSKM) is a key part of sustaining a high level of process safety performance. Effective PSKM is ethically correct and financially attractive. A well-defined and implemented Process Safety Knowledge Management program can lead to improved occupational and process safety, reduced environmental impacts within an organization, and business continuity. There is also a financial benefit to having an effective PSKM System. Reduced costs due to implicit and explicit financial impacts are realized when businesses run efficiently. Having an effective PSKM System in a company provides agility and flexibility to market changes due to reductions in process safety incidents, reduced production downtime, and decreased loss of business opportunities. Examples of financial and business benefits for PSKM include:

1. Reduced costs associated with Process Hazard Analysis (PHA):

    Knowledge from the scenarios developed during the PHAs is sustained instead of having to be constantly recreated. AMOCO Oil found in the early 1990s that doing multiple PHAs (3 to 4) during a project, versus just one PHA/HAZOP at the design phase, saved enough money in start-up costs to pay for all project PHAs several times over [13].

2. Minimized chance for mistakes made by a new operator leading to reduced downtime and fewer process upsets:

    Chevron's energy-use network generated an initial $150 million in savings in its first year with a total over time of $650 million between 1991 to 1999 from just one community effort. Across Chevron, from 1992 to 1999, productivity increased 30 percent and employee safety performance improved 50 percent. "Of all the initiatives we have undertaken at Chevron during the 1990s, few have been as important or as rewarding as our efforts to build a learning organization by sharing and managing knowledge throughout our company. In fact, I believe this priority was one of the keys to reducing our operating costs by more than $2 billion per year from about $9.4 billion to $7.4 billion over the last seven years" said Kenneth Derr, who retired as the Chairman of the Board of Chevron in 1999 [14].

3. Improved use of lessons learned:

    Use of lessons learned minimizes overlooking of potential safety hazards during the work permit approval process for maintenance or shutdown activities or when the unit is in restart especially in cases where similar incidents have happened before.

4. Improved plant reliability and reduced downtime:

    At Schlumberger, the InTouch system (a technical support service to field operations) created a centralized knowledge-based organization, with easy access to information. The results were $150 million cost savings a year, a 95% reduction in time to resolve technical queries and a 75 percent reduction in time to update engineering modifications [14].

5. Improved decision-making and reduced errors while responding to abnormal situations [15]:

    Brewing giant Anheuser-Busch, the maker of Budweiser, responded to the COVID-19 pandemic and sanitizer shortage by repurposing facilities in Van Nuys, California and Baldwinsville, NY (both USA) to produce and distribute hand sanitizer [15].

    Ford Motor Company partnered with 3M to produce air-purifying respirators for health care workers on its auto assembly lines in Michigan (USA) in response to the COVID-19 pandemic [16].

6. An effective PSKM system enables Corporate Responsibility, Business Flexibility, Risk Reduction, and Sustained Value which were identified as the four benefits in the CCPS study "The Business Case for Process Safety" [12]:

    The chemical processing and petroleum companies that participated in the CCPS study, combined with data from other sources, report the following returns from their investment in process safety [12]: increased productivity (up to 5% increases in productivity, due mainly to increased reliability of equipment), increased revenues of $50 million, decreased production costs (up to a 3% reduction in production costs resulting in a savings of $30 million), decreased maintenance costs (up to a 5% reduction in maintenance costs resulting in savings of $50 million), required lower capital budget (up to a 1% reduction in capital budget resulting in savings of $12 million), and lower insurance premiums (up to a 20% reduction in insurance costs resulting in savings of $6 million).

An effective PSKM system generally includes the following three phases:

- Phase 1: Generating Process Safety Knowledge
- Phase 2: Retaining Process Safety Knowledge
- Phase 3: Sharing Process Safety Knowledge.

2.1   Generating Process Safety Knowledge

Generated knowledge needs to be immediately classified, stored, and maintained. Classification should be done based on some taxonomy and should be based on the company's management structure [17]. The process of generating Process Safety Knowledge within an organization can be used for problem-solving: to resolve process errors, to address abnormal situations, to improve decision-making, and to ensure that the workforce is trained and competent.

Effective management requires pertinent data-driven information for decision-making. Most organizations today have process safety data and information. The data is scattered in various procedures, Safety Data Sheets (SDS), technical documents, Systems of Record (SOR, information storage and retrieval systems that store important data on a system or process within an organization) and manuals. However, having the data by itself is insufficient. It should be aggregated and analyzed to create usable information and knowledge as it forms a critical foundation for making informed decisions.

Although data can be sorted, statistically characterized, and reported, the knowledge to be gained requires the use of skills, experience, and ability to extract its true meaning or to generate the wisdom to effectively influence others. To develop Process Safety Knowledge that can be operated upon (operative Process Safety Knowledge), even though the data and information may exist, it is fundamental that personnel involved know when and how to use it and equally important, how this data and information is related to their responsibilities.

Operative Process Safety Knowledge can lead to:

- Ability to recognize hazards
- Preventing process safety incidents to achieve CCPS vision of a "World without process safety incidents"
- Preventing non-compliance
- Preventing rework
- Continued process safety improvement

PHAs conducted in an organization can be examples of effective knowledge management, as the hazards and controls identified during the PHA must be communicated to the affected personnel such as operations and maintenance. When operators are provided with clear communication about identified risks, they can utilize the information generated by the PHA and act on it. This concept is the transition from available information to knowledge. A thorough understanding of the process and its risks leads to the right decisions being made during processing process startup or abnormal situation management.

Process Startup is one of the operational phases reviewed during PHA. An example of a startup hazard that was not thoroughly understood was the 2005 BP Texas City incident [reviewed in Chapter 6]. The incident occurred during the start-up of an isomerization (ISOM) unit when a raffinate splitter tower was overfilled and pressure relief devices opened resulting in a flammable liquid release from a blowdown stack that was open to the atmosphere. The release of flammables led to an explosion and fire resulting in 15 fatalities. A key factor in this incident was that during the startup after a maintenance outage, operations personnel pumped flammable liquid hydrocarbons into the tower for over three hours without any liquid being removed, which was contrary to the startup procedure instructions. Defective critical alarms and control instrumentation provided false indications that failed to alert the operators of the high level in the tower. A deeper knowledge and understanding of the process by the operators could have alerted operations that the startup was not proceeding normally.

An organization with immature PSKM could conduct risk assessments with a team lacking the right expertise and experience. This situation may lead to a poor-quality assessment. Useful risk assessments promote knowledge of the hazards and effective decision-making in case of deviations from normal operations. A risk assessment conducted by an experienced team can be completed on time because of the availability of crucial information and sharing of knowledge. As a result, the organization can realize financial benefits from a study that is completed with a high level of technical competence. More importantly, the organization understands the risks and benefits from a lower probability of repeat incidents. The organization grows and gains wisdom.

An organization having operative Process Safety Knowledge will not have to over-rely on outside parties for process safety and equipment reliability competency as sufficient knowledge has been generated in-house. An organization that over-relies on outside parties for such competencies may not be generating sufficient knowledge in-house. This dependency could cause the organization to divert its attention from certain elements of RBPS such as risk assessments, asset integrity, and incident investigations. For example, outsourcing of equipment maintenance may lead to a failure to generate an understanding of why the equipment fails. In many such situations, operators may know how to operate the equipment but may have only limited information or knowledge about troubleshooting failures or operating deviations of the equipment. The lack of knowledge could lead to potentially unsafe choices when operators, who do not understand the process, troubleshoot issues they do not understand well. A better practice, in this case, would be to invest in training from the service provider, to ensure all operational personnel have a baseline understanding of the equipment maintenance and troubleshooting requirements. In-house baseline operational and maintenance knowledge would reduce overall downtime through greater control of maintenance and repair schedules. Ultimately it will improve in-house knowledge and enhance the safety of troubleshooting.

Having knowledge is especially important in emergency response. This knowledge can be generated as part of the execution of the program. Most companies provide information regarding the hazardous chemicals on-site to first responders in the form of SDS and site maps as part of their stakeholder outreach. Ideally, relevant safety information should be converted into knowledge by sharing how to respond to certain scenarios by developing pre-plans and performing in-person emergency drills. This way responders have a better understanding of how to respond and what to expect in case of an incident [7]. A best practice is to invite the first responders to the facility and provide a walk-through of the facility on a regularly scheduled basis. This facility walk-through covers the location of major hazardous material inventories, on-site response equipment, and potential access challenges. Many facilities perform a portion of their drills with first responders. Drills and collaborative practice should reduce the requirement for this information to be conveyed during incidents when time is critical. Another advantage that can be realized is keeping relationships and lines of communication between the responding team and site emergency planners open.

Not generating PSK ends up costing an organization. Failing to effectively close action items, not sharing Process Safety Knowledge between RBPS elements, or generating inaccurate PSK information are all examples of these types of failures that can add costs to an organization.

An organization that has a process safety management system generates various action items through PHAs, Audits, MOCs, Capital Projects, and Incident Investigations. These action items help the organization to rectify any deficiencies and further generate new or correct data, information, and knowledge (tacit and explicit). Actions should be assigned to team members and prioritized to rank which actions need to be closed first. Resources may have to be allocated to make sure deficiencies are corrected in a timely manner. Closing actions based on a risk-based approach to resource priority improves the organization's performance with respect to risk management. In addition, actions should follow the SMART model which means that they should be written with Specific, Measurable, Attainable, Realistic, and Timely requirements and goals. The SMART method provides clarity and focus of efforts for productive and timely resolution. The SMART method should also include the basis or reason for the action to capture its intent and ensure knowledge is not lost in the process.

RBPS elements, by themselves or in combination, generate PSK to identify systemic issues and interrelationships between elements. For example, the facility emergency plan may identify a scenario that requires an Io. Such information should be discussed in the PHA as a high-severity or high-consequence scenario. Worst case or alternate case scenarios identified in an organization's Risk Management Plan can be safety-critical scenarios in the operating procedures. Each element of the safety management system should complement, not duplicate, information in other elements to ensure clear dissemination of information. Such a practice would prevent duplication of work while disseminating the knowledge of specific process safety risks.

"Inaccurate" knowledge could be generated if PSI is not kept vetted, current, accurate, and accessible. "Inaccurate" could mean any of the following: incorrect, incomplete, out-of-date, inconsistent, undocumented, unofficial, "always done that way", unverified, and useful knowledge used in the wrong place/time (e.g., hammer used for a screw). Inaccurate knowledge may be much worse than lack of knowledge. For example, if corporate or facility policies are not being followed or they are ignored, inaccurate knowledge might be generated. The divergence between what an organization knows (i.e., knowledge) and what the reality is (i.e., truth) leads to generating "inaccurate" knowledge. An example is the T2 Laboratories event when T2 was unaware of a second exothermic reaction that occurred in the batch recipe at elevated temperatures. The relief system was designed for normal operating conditions and was unable to properly protect the vessel during a reactive relief event. The relief system would relieve; however, the relief was not sufficient to prevent the vessel failure [18].

2.2   Retaining Process Safety Knowledge

When an experienced, specialized, or highly knowledgeable individual leaves a critical position due to retirement, illness/death, or reassignment, multiple challenges can arise for the organization that is tasked with backfilling that position.

Where loss of knowledge in an organization becomes even more problematic is in the matter of knowledge retention. If the organization has documented the tasks or procedures related to job responsibilities, there is a greater chance of an uneventful transition. Otherwise, the incoming employees will be vulnerable to hazards that they do not see, but that their predecessor understands well. The realization of those hazards could lead to a range of safety or quality impacts. Those impacts could be minor, such as first aid or minor injuries, or potentially could become major, causing fatalities. Quality effects could be unfavorable to an organization's profitability if the new candidates lack sufficient training or experience, thus affecting operational efficiency. A Management of Organizational Change (MOOC) program can be used to address the potential issues described above. For more information on the importance of organizational change management, refer to the Center for Chemical Process Safety (CCPS) book *Guidelines for Managing Process Safety Risks During Organizational Change.*

The 1974 Flixborough incident served as an example of how a failure to retain knowledge led to a significant event. The incident resulted in 28 fatalities onsite, 53 offsite injuries, 1,800 nearby houses damaged in the area beyond the plant fence line, and property damage exceeding $425MM ($2.3 Billion adjusted for inflation). The incident was a result of the failure of a temporary bypass around a cyclohexane reactor. A critical contributing cause to the event was the absence of an experienced maintenance engineer on site due to the departure of the site maintenance engineer. As a result, both process information and process knowledge were not available to ensure a suitable engineered solution. Consequently, stress analyses calculations were not performed on the bypass connection which contributed to the release and eventual ignition of cyclohexane.

This example presents the dangers associated with process and equipment hazards that are not documented or recognized. When critical steps of a procedure or a task are not documented, it could lead to potential injury since key factors may be missed. Additionally, the same effect is signaled when key personnel in positions of knowledge management are absent. Organizations that effectively use process knowledge and manage the Process Safety Knowledge can reduce the likelihood of adverse effects of hazards. If an organization uses their PSKM systems instead of relying on a limited number of employees to keep this knowledge, the organization has a higher probability of having correct and current PSK readily available for transfer to new employees, or employees who are changing jobs within the company.

An organization needs to problem-solve at all levels of the organization and not have the "hurry and get it fixed, we'll teach you later" mentality. This can become a daily rhythm when the company practices mentorship or coaching programs so that a person without the knowledge actively engages and learns from a more experienced person. A recommended practice is to adopt the Management of Organizational Change (MOOC) when changing safety-critical roles, given that the MOOC also ensures that PSK is transferred, distributed to others, or that the risks of new personnel cannot effectively access the PSK is otherwise mitigated.

2.3   Sharing Process Safety Knowledge

Sharing Process Safety Knowledge in the form of lessons learned [19] helps to eliminate repeat incidents and reduces the potential to duplicate inaccurate knowledge and work. Experiencing repeat incidents is an indicator that the PSKM system is inadequate, and that corporate learning is compromised.

Repeat incidents often arise because underlying and latent causes are not properly identified or addressed. Management system failures that are latent causes are often widespread within a company and when recognized bring multiple opportunities to share lessons learned. When that sharing does not occur, the latent cause can remain in place at some points within the organization, even though it has been recognized and rectified in others. The 2005 BP Texas City incident (which led to 15 fatalities and over 180 injuries) serves as an example of inadequate sharing of information. The blowdown drums were known to be a problem and for this reason, BP had standardized other systems that were not yet adopted at BP Texas City. The blowdown system at the BP Texas City was originally installed in the 1950s to safely contain liquids and combust flammable vapors released from the process. However, it was never connected to a flare system [20]. The way to remember those lessons is by making them vivid when training, incorporating them in our technology and daily operations, and periodically reinforcing them [19].

Habits, good or poor, are learned behaviors that tend to stick when formed. A good habit results in a positive outcome. In the context of Process Safety Knowledge, a good habit generates good and useful Process Safety Knowledge and will be repeated if there is a recognized benefit. A bad habit results in a negative outcome and, in the context of Process Safety Knowledge, discourages useful Process Safety Knowledge. Poor safety culture helps the formation of bad habits. Without a clear vision of the benefits and rewards of good Process Safety Knowledge Management and the necessary planning needed to accomplish it supported by continuous training, good habits will not be developed. The costs of not reproducing good Process Safety Knowledge habits will result in operational deficiencies.

Inaccurate or incorrect knowledge, in the context of Process Safety Knowledge Management, is knowledge that is not aligned with the company's policies and other rule-based facts or requirements. Inaccurate or incorrect knowledge is reproduced when it is not recognized as such. Generally, reproducing inaccurate or incorrect PSK multiplies the cost of the initial poor knowledge that has been reproduced, often without an awareness of the impact. As an example, if the root cause is not identified correctly, inaccurate, or incorrect knowledge can be generated and shared. Sharing inaccurate or incorrect knowledge may be riskier than having a lack of knowledge.

Rework occurs when knowledge is not shared because roles are not clearly defined or communicated. In each company, subject matter experts (SMEs, often as experienced personnel well-vetted contractors) are present. However, when there is no clear communication about who these subject matter experts are, rework and productivity losses may occur. When there is deliberate mentoring and competency development and the SMEs are known, problems can be fixed faster, reducing time spent trying to resolve issues amongst teams that lack the required skillset. Examples of rework would be costs associated with the correction of deficiencies, defects, or non-conformances.

Sharing Process Safety Knowledge promotes optimal solutions. This can lead to the development of best practices within an organization, or across similar industries. The Process Safety Beacon® produced monthly by CCPS for plant operators and manufacturing personnel around the world is an excellent example of industry knowledge sharing. Each issue presents a real-life incident, and describes the lessons learned and practical means to prevent a similar incident.

Problem solving and innovation are encouraged when knowledge sharing occurs. When knowledge is shared, employees are encouraged and motivated to problem solve on their own, applying that shared knowledge to their respective problems and issues. This has the effect of nourishing morale plus it challenges employees to collaborate to jointly solve problems.

Where knowledge is not shared, and where knowledge retires with the personnel who hold it, there is a tendency for the work that depends upon that knowledge to be exported out of the company or siloed within the company. An example of siloed knowledge is when the retiree is rehired due to the lack of process knowledge sharing and the retiree once again holds the information closely. <u>Organizations must create and implement a Process Safety Knowledge transfer process to store and share critical information with future teams</u>. This practice will reduce costs associated with employee transfers and onboarding time for new hires. It will also ensure critical Process Safety Knowledge is stored and shared with future team members.

2.4   From Knowledge to Wisdom

Process Safety Knowledge, and by extension Process Safety Knowledge Management, must be present to achieve Process Safety Wisdom. Process Safety Wisdom is what allows individuals and groups to move beyond compliance with standards and even currently considered best practices, and to develop new best practices that can be shared throughout all industries to reduce and eliminate process safety incidents.

One of the key differences between knowledge and wisdom is taking the proper action, when and where needed. Although the information and knowledge available may not change, increased wisdom may lead to new means of using that information for improving the process. One of the most common ways of advancing process safety wisdom is by finding new ways of using existing information to identify new corrective actions. Analysis techniques such as LOPA (Layer of Protection Analysis), Bow Tie Analysis, and Facility Siting Methodologies are the result of Process Safety Wisdom furthering the development of new methods of identifying previously under-reviewed or unknown process safety gaps.

Much of the Process Safety Wisdom that has been obtained previously was obtained through a review of previous incidents and case studies. These events gathered attention, and through their investigation and understanding some wisdom was gained, but in many cases at the cost of lives. Through Process Safety Knowledge Management, Process Safety Wisdom is best obtained from documenting, teaching, and training on existing process knowledge and lessons learned by SMEs willing to increase knowledge sharing expectations and the wisdom of others.

3   PSKM and Risk Based Process Safety

"Of central importance is the changing nature of competitive advantage - not based on market position, size and power as in times past, but on the incorporation of knowledge into all of an organization's activities."
Leif Edvinsson, Swedish Intellectual Capital leader in Corporate Longitude [21]

Risk Based Process Safety (RBPS) is a framework of documented management systems and activities to help organizations design and implement more effective process safety programs. RBPS incorporates four pillars:

1) Commit to Process Safety

2) Understand Hazards and Risk

3) Manage Risks

4) Learn from Experience

The Center for Chemical Process Safety (CCPS) thoroughly describes these pillars and associated 20 elements in its publication, Guidelines for Risk Based Process Safety (RBPS) [7]. One of the elements under the second RBPS pillar (Understand Hazards and Risk) is Process Knowledge Management. Process Safety Knowledge Management (PSKM) is different from the Process Knowledge Management (PKM) element. It directly impacts the four pillars of RBPS while operating within and beyond the limits of the individual elements of each pillar. In fact, PSKM is an expansion of PKM whereby it is actively used and maintained. PSKM is not just a catalog of information, but instead knowledge that is used to safely manage a process.

The intent of this chapter is to:

- Describe the differences between the RBPS element on Process Knowledge Management and Process Safety Knowledge Management
- Demonstrate how PSKM directly impacts the four pillars of RBPS
- Provide case study examples of PSKM with a RBPS element while extending beyond that element

3.1   Process Knowledge Management vs. PSKM

Process Knowledge Management (PKM) is an element of the second pillar of RBPS: Understand Hazards and Risk. The Process Knowledge element refers to the collection of process safety information, such as the information pertaining to the process chemicals (hazards of the process), process technology, and process equipment. It involves work activities associated with compiling, cataloging, and making specific sets of recorded data available. Examples of these work activities are found in the Example Inputs and Outputs of the DIKW Pyramid in Chapter 1, Table 1-1.

Process Safety Knowledge Management (PSKM) is defined as a system for capturing, organizing, maintaining, and providing the right Process Safety Knowledge to the right people at the right time to improve process safety in an organization. PSKM extends PKM by combining Process Safety Knowledge (PSK) with a Knowledge Management (KM) system. It governs how an organization continuously manages and improves PSK such that it can be applied to the process by everyone at all levels of the organization. This is how an organization creates knowledgeable and competent people to make the right decisions when they are needed. This PSKM System is represented in Figure 3-1 as a funnel to capture all the knowledge and information from those in various roles and responsibilities and with different backgrounds collaborating and working together.

The Capture and Organize stages for PSKM are very similar, if not identical, to Process Knowledge Management, as described in the RBPS book [7]. As the knowledge is organized, it becomes refined and targeted. The Maintain and Provide stages of PSKM enable an organization to use this knowledge effectively. For example, having the right Safety Data Sheet (SDS) is required, but it is not sufficient. The information in the SDS should be extracted and maintained as useful knowledge. The maintenance of this knowledge might be managed by a smaller group. The delivery of the knowledge needs to be targeted to specific people when it is needed. For example, companies in the process industries that manage hazardous chemicals should have Safety Data Sheets for those chemicals as part of their Capture activities. However, in an emergency, the company needs to be able to provide knowledge to emergency response teams

The PSKM Triangle is an inverted triangle that shows knowledge funneled, refined, and provided to the end-user.

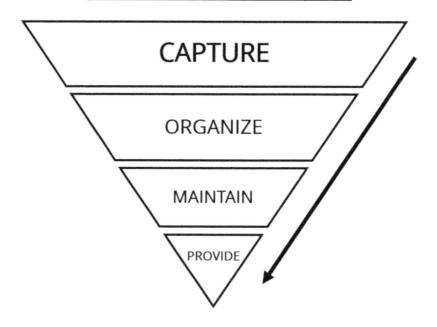

Figure 3-1 PSKM system

that goes beyond the information in the Safety Data Sheets. This knowledge transfer would include all hazards detailed in the Safety Data Sheets, along with locations of inventories, identification of process equipment that contains the hazardous materials, whether adjacent units should be shut down and whether the people in adjacent units should evacuate or shelter in place. Knowledge goes beyond data and could for example lead to providing P&ID-based procedures and methods to isolate, neutralize or dispose of hazardous materials, rather than simply providing the hazards as listed in the Safety Data Sheet. Chapter 4 provides a detailed discussion of the PSKM System implementation.

As introduced in Chapter 1, data is individual descriptive pieces of information, or parameters. Information is the organization of data and its relationship to other parameters. Knowledge is understanding of information through training, and wisdom is experience gained by applying knowledge to the process and the ability to use that knowledge to change the process. In a PSKM program, the goal of training is to share data and information with those being trained, combined with the competence and experience of the trainers. In this way those being trained gain a deeper understanding of the information shared, which is the beginning of knowledge for them. Knowledge leads to being able to use information to predict what could happen next during operation or as an incident is developing. Experience comes with the application of knowledge. Experience is gained through time, exposure, and practice. One can build experience in a variety of ways including through simulations and tabletop exercises as well as through being involved in emergency situations and incidents. Experience guides actions that lead to wisdom. As events occur, wisdom, which is the ability to apply what was previously learned to new situations, guides the actions needed to respond effectively. These events bring change. Subsequently, change results in new data/information generation and the Data to Wisdom cycle repeats itself. The Continuous PSKM Cycle is represented in Figure 3-2.

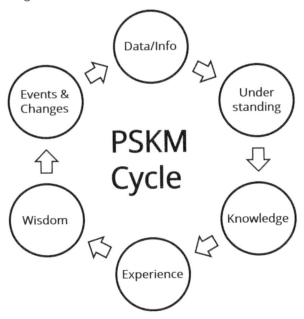

Figure 3-2 Continuous PSKM cycle

PSKM is a continuous improvement process. Functionally, the PSKM System needs to be able to capture, organize, maintain, and provide the knowledge in the organization. As the data is collected into information, and information is turned into knowledge and subsequently wisdom, it will follow a continuous improvement process. In the Continuous Improvement process in Figure 3-2, a progression from one element of the DIKW pyramid to the next element is shown. For example, moving from Data/Information to Knowledge requires Understanding. Moving from Knowledge to Wisdom requires Experience. The cycle is completed when Events or Changes are incorporated which result in the generation of new Data or Information.

While a connection can be made from each of the 20 RBPS elements to the PSKM cycle, the four most directly connected RBPS elements to the PSKM cycle are Process Knowledge Management, Operating Procedures, Training, and Management of Change. Knowledge is generated in Process Knowledge Management, as described above. Knowledge is shared in Operating Procedures and Training. Knowledge is also shared through Hazard Identification and Risk Analysis (HIRA), as teams and individuals evaluate the existing safeguards and Independent Protection Layers (IPLs) to determine if there are gaps or deficiencies as compared to company risk criteria. Operating Procedures define Safe Operating Limits, but do not contain design basis information for safety relief devices nor do they contain verification details for the adequacy of Safety Instrumented Systems (SIS). These examples of knowledge are part of the HIRA process and are typically found in relief system design basis documents and process control narratives, respectively. Finally, knowledge is updated using the Management of Change (MOC) process. The MOC process requires an update to Process Safety Information (PSI) related to the managed change, and PSK is developed through the validation of the information. Although only these four elements are mentioned here, all elements and pillars within RBPS interact with PSKM directly or indirectly. This chapter will highlight the most relevant elements in each pillar even though all RBPS elements play an important role in the effectiveness of the PSKM System.

The PSKM Cycle impacts the four pillars of RBPS as it operates on the elements within each pillar as shown in Figure 3-3. The PSKM Continuous Improvement Cycle occurs in each element of RBPS. As the organization commits to process safety, the PSKM Cycle enables the commitment to process safety to mature and grow by moving information to knowledge and finally to wisdom. As each of the four pillars of RBPS evolves such that the commitment to process safety enables understanding the hazards and risk, management of risk and learning from experience, PSKM enables organizational maturity by moving into wisdom in each pillar of RBPS. Once learning from experience is institutionalized, it further strengthens the organization's commitment to process safety, hence advancing process safety maturity in the whole organization. Chapter 4 will introduce how to build a PSKM System that will improve organizational wisdom, and Chapter 5 will present how to audit and improve such a PSKM System to make it sustainable.

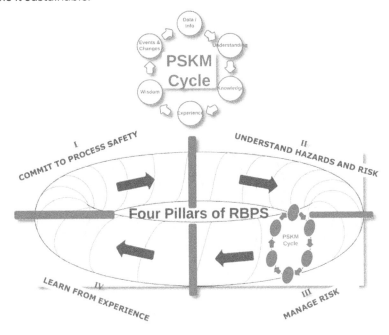

Figure 3-3 Impact of the PSKM Cycle on the Four RBPS Pillars

3.2   PSKM Cycle Impacts on the Four Pillars of RBPS

The RBPS Pillars and associated elements [7] are presented in Table 3-1. Elements directly impacted by Process Safety Knowledge Management are highlighted in bold in Table 3-1 and covered in the following sections.

3.3   Commit to Process Safety

The first pillar under RBPS is Commit to Process Safety. This pillar is the foundation of Risk Based Process Safety, and it covers the following five elements: 1) Process Safety Culture; 2) Compliance with Standards; 3) Process Safety Competency; 4) Workforce Involvement; and 5) Stakeholder Outreach

Process Safety Culture, Process Safety Competency and Workforce Involvement will be discussed in detail in the following subsections.

Table 3-1 RBPS Pillars and Corresponding Elements

Pillar	Element
Commit to Process Safety	1. **Process Safety Culture**
	2. Compliance with Standards
	3. **Process Safety Competency**
	4. **Workforce Involvement**
	5. Stakeholder Outreach
Understand Hazards and Risk	6. **Process Knowledge Management**
	7. **Hazard Identification and Risk Analysis**
Manage Risk	8. **Operating Procedures**
	9. Safe Work Practices
	10. Asset Integrity and Reliability
	11. Contractor Management
	12. **Training and Performance Assurance**
	13. **Management of Change**
	14. Operational Readiness
	15. Conduct of Operations
	16. Emergency Management
Learn from Experience	17. **Incident Investigation**
	18. **Measurement and Metrics**
	19. **Auditing**
	20. **Management Review and Continuous Improvement**

(Note: Elements in **Bold** are directly impacted by PSKM.)

3.3.1   Process Safety Culture

Process Safety Culture defines the values and behaviors of the organization including the organization's attitude towards Process Safety Knowledge Management. Process Safety Culture (PSC) promotes the development, implementation, maintenance, and monitoring of an effective PSKM system in the organization by:

- Ensuring high performance standards are established, enforced, and understood by the workforce.
- Establishing an environment for the sharing of knowledge by training and cooperation.
- Ensuring that continuous monitoring of performance is available, current, and accurate.
- Sharing and incorporating events that result in changes to the knowledge in a timely manner.

In return, PSKM supports PSC by providing information about the organization's process safety culture, core values and mission statements. Without a healthy and mature process safety culture throughout all elements of RBPS, PSKM will struggle to thrive. A typical example of this phenomenon is an organization that concentrates on experience without developing the necessary wisdom through root cause analysis and as a result the organization does not make it through the full PSKM cycle as shown in Figure 3-2. Through a strong PSKM System, an organization can learn to generalize and deal with an unforeseen or unfamiliar hazard.

Michele Gelfand, a distinguished professor from the University of Maryland, College Park, published a book entitled "Rule Makers, Rule Breakers: Tight and Loose Cultures and the Secret Signals that Direct our Lives" in 2018 [22]. In this book, she provides a framework for analyzing how behavior is highly influenced by the perception of threats. In fact, she points out the tight-loose ambidexterity within organizations. An organization can become tight (follow orders) when faced with a threat such as dealing with a pandemic but also loosen up (break rules) when trying to innovate. A strong PSKM system enables an organization to tighten up when faced with the threat of potential loss of containment of a highly hazardous chemical. A strong PSKM System also allows for the organization to become innovative when dealing with a new or novel hazard. In that sense, PSKM provides the necessary tools and methodologies that an organization must possess to have a strong Process Safety Culture.

3.3.2   Process Safety Competency

Process Safety Competency allows an understanding for turning information into knowledge hence generating and sharing the knowledge within the organization. Process Safety Competency ensures that the necessary skills and understanding are available to manage process safety risk and it has three main concepts at its core:

- Ensuring appropriate knowledge is available, deliverable, current, and accurate.
- Improving skills and knowledge by training.
- Applying the knowledge that has been learned to promote risk-informed decision making.

A competent employee is knowledgeable and can apply that knowledge. Process Safety Competency, however, is not just about having competent employees but instead is about how to determine and define what competency looks like and laying out a plan/program to create a competent people, and then maintaining competency over time in an ever-evolving manner. If the organization is not clear about what competency is and how to get its employees to acquire/maintain competency, it will not have a competent workforce. The organization should strive to create a competent workforce in addition to just trying to hire competent people, given that the organization understands that training is not the same as competency and that highly trained people may also need guided experience to become competent.

The competency element drives the PSKM so that the organization's employees can apply what they learn. It involves work activities that promote personal and organizational learning and helps ensure that the organization retains critical knowledge in its collective memory. The competency element provides new information for the learner; however, knowledge provides a means to catalog and store information so that it can be retrieved on demand.

Knowledge that has a decisive or crucial importance to success or failure of processes is defined here as "Critical knowledge". It is gained through real-life experiences inside an organization as well as through shared experiences from industry. PSKM enables capturing, categorizing, maintaining, and delivering this critical knowledge. It is the Process Safety Knowledge Management that contributes directly to outcomes while preventing losses, closing knowledge gaps, and supporting predictable operations. comp can be centralized or decentralized. In a centralized structure, critical knowledge resides with a single team member or unit. In a decentralized structure, critical knowledge is shared across the organization with all employees. Having a strong PSKM System in an organization makes it easier to have a decentralized structure. Process Safety Culture should also support a decentralized structure incentivizing knowledge sharing among all employees. PSKM will also be critical in accessing information outside of the organization through membership-based consortiums such as Center for Chemical Process Safety (CCPS), American Petroleum Institute (API), Process Industry Practices (PIP), Environmental Technology Council (ETC), and others.

Examples of work activities that help ensure the organization retains critical knowledge by implementing a strong PSKM system are listed here:

- Maintaining P&ID Revision History that provides information about equipment or instrumentation changes made to the process over time.
- Creating competency matrices that define roles and responsibilities, competency requirements and qualifications needed for greater levels of responsibility.
- Developing and maintaining accurate Operating Procedures that clearly describe step-by-step activities along with checklists that are easy to understand and implement and validating those procedures with conceptual and field exercises.
- Training that includes knowledge transfer related to Chemical Reactivity within a process or unit operation. Chemical Reactivity knowledge can be obtained from the Chemical Reactivity Worksheet [18, which is available from CCPS.
- Recommendations made during hazard review studies or audits and the respective closure documentation that provides information about the measures taken to address these recommendations.

Process Safety Competency, when integrated with PSKM, supports almost all other RBPS elements. Examples of this support are presented in Table 3-2.

Table 3-2 Process Safety Competency Interactions with RBPS and PSKM

RBPS Element Supported	PSKM Example Interactions
Training and Performance Assurance	• Understanding of job and task requirements
Compliance with Standards	• Understanding of local applicable rules and regulations, RAGAGEP and best practices
Hazard Identification and Risk Analysis (HIRA)	• Understanding of process safety hazards, subsequent potential consequences, and barriers
Operating Procedures	• Understanding of the written instructions for safe manufacturing • Understanding of potential consequences of deviation and steps to correct any deviations
Incident Investigation	• Understanding of the process to report incidents • Knowledge of root cause analysis • Experience to investigate incident trends • Sharing learnings from incidents across the enterprise, i.e., learning from others

3.3.3   Workforce Involvement

Workforce Involvement provides a system for enabling the active participation of company and contractor workers in maintaining and delivering the knowledge within the organization through various means such as standard operating procedures, training, risk assessments, and lessons learned. PSKM has a strong connection to Workforce Involvement which, in turn, supports the PSKM cycle. A healthy Workforce Involvement promotes active participation of company and contractor workers at all levels within the organization and has four core concepts relevant to a successful implementation of PSKM:

- Maintaining a Dependable Practice by improving skills to involve competent personnel to guide actions or share knowledge.
- Conducting Work Activities so that all personnel can share information by providing input or feedback to the organization.

- Monitoring the System for Effectiveness by developing metrics that provide information and opportunities for improvement.
- Actively Promoting the Workforce Involvement Program through training and providing opportunities for employees to grow in the areas of process safety improvements.

As explained in the previous section, Process Safety Competency defines roles and responsibilities through Competency Matrices. Workforce Involvement ensures that employees at all levels of the organization have their competency/knowledge needs identified and their thoughts and ideas for improving PSKM are actively solicited. Through this element, company as well as contractor workers at all levels in the organization can take actions to improve training, knowledge capturing, and other essential aspects of PSKM. Having Workforce Involvement means that information flows freely throughout the organization rather than simply in a top-down manner, allowing every member of the organization to contribute to knowledge capturing, sharing and its application. PSKM further provides a system for design, development, implementation, and continuous improvement of the RBPS management system.

Ownership of the PSKM system within the organization is an activity for which roles and responsibilities should be assigned. PSKM responsibilities may be assigned to a single individual, but they are usually distributed among many different people from diverse disciplines. For example, a technical manager might be responsible for keeping multiple types of process safety information current (e.g., P&IDs, equipment files), but the activities will be distributed among drafters, process engineers, project engineers, and others. Different aspects of PSKM are provided by different people within the organization, such that no one person has all the knowledge. Maintenance people provide data on how controls are maintained and about frequent failures that occur in the field. Engineers provide data on design and function of the controls. Operations provide data on how the controls are utilized, gaps/opportunities and needs of the process. Ultimately, each member of the workforce "owns" data and the corresponding knowledge that they contribute to a successful PSKM program. For example, the maintenance department may own the technical information in the Original Equipment Manufacturer (OEM) manuals and understand equipment constraints. Similarly, the operations department may use this information to develop operating limits and consequences of deviation. With this approach, the responsibility of sustaining Process Safety Knowledge is widely understood throughout the organization. Both groups would then be able to share knowledge if there are any opposing goals between ideal process conditions and longevity of the equipment's life.

Organizations should decide who will be responsible, accountable, consulted and informed of the various elements of PSKM. This concept of PSKM activities and decision-making authorities can be defined by a RACI (Responsibilities, Accountabilities, Consulted, Informed) chart. Chapter 4 presents an example of a PSKM RACI Chart.

3.3.4 Case Study – CCPS Member Company - Process Safety Knowledge Ownership

A CCPS Member company (referred to here as the Member Company) recognized that there was a significant gap in the company's Process Safety Knowledge. The process safety leader learned through participation in process hazard analyses that he was the only person within the company's process group who had the experience and knowledge specific to the process being studied that could provide cause and consequence information. Process engineers were not competent in process safety, nor did they recognize that they were responsible for process safety. The Member Company subsequently learned that competency requirements for chemical engineers varied globally due to differing chemical engineering curriculums. Europe's Chemical Engineering curricula were chemistry based; China's front-end design based, and US/UK based on foundations of basic sciences. Process safety was not a focus of nor included in any of the chemical engineering curricula. The Member Company soon developed competency requirements for chemical engineers that included process safety. The Member Company established Process Safety Knowledge Owners (PSK Owners) for each of their product lines consisting of a process engineer and a chemist as co-owners. Adding Process Safety Knowledge Owners brought value to the employees. Employees understood their roles in process safety and were able to contribute to an effective program bringing additional value and diversity to their job function. This career path had the positive effect of reducing employee turnover as they sought additional job development/growth opportunities. In addition, the Member Company divided PSK Owners into three levels with Level 1 being recent university graduates, Level 2 individuals having 3-5 years of process safety experience, and level 3 individuals with 5-10 years of process safety experience. These tiers allowed for progression of employees, expansion of knowledge over time and a formal way for the company to develop "experts" for each of their product lines. The

Member Company's training program shifted from half-day informal meetings to formal training followed by test verification for those in safety critical roles such as PSK Owners. Training was transformed from one-way communication through computer-based delivery to in-person learning to allow trainees to interact with the trainer by asking questions and receiving answers in a two-way communication format. By having a structured program for process engineers to develop their PS knowledge, the Member Company has seen an increase in retention, decrease in time taken to perform PHA revalidation and better availability of key safety information. The program also highlighted the need for a similar approach across manufacturing for people with a role in process safety.

3.4 Understand Hazards and Risk

The second pillar under RBPS is Understand Hazards and Risk. PSKM has a strong tie to Understand Hazards and Risk by sharing potential hazard information to gain knowledge. This pillar covers:

1. Process Knowledge Management

2. Hazard Identification and Risk Analysis

3.4.1 Process Knowledge Management and Process Safety Knowledge Management

Process Knowledge Management deals with developing, documenting, and maintaining process knowledge. The Process Knowledge element primarily focuses on information that can easily be recorded in documents, such as (1) written technical documents, and design standards and specifications, (2) engineering drawings and calculations, (3) specifications for design, fabrication, installation of process equipment, and (4) other written documents such as safety data sheets (SDSs) [23]. This collection of information is referred to as the Process Knowledge. However, knowledge includes understanding the information. In that respect, the competency element is required to initiate the transition of the body of collected information into knowledge. Process Safety Knowledge Management extends this approach beyond the information that can easily be recorded in documents. Once implemented as a PSKM system, it involves every element of RBPS and extends knowledge to wisdom through experience (either acquired first-hand or through external learning and simulations). The implemented PSKM System also incorporates providing knowledge at the right time to the right person in the organization.

3.4.2   Hazard Identification and Risk Analysis

Understanding risk is essential to the entire concept of RBPS. Hazard Identification and Risk Analysis (HIRA) is the process of identifying hazards and evaluating risk in the workplace. This process is achieved through:

- Conducting observations and field audits.
- Conducting process safety studies.
- Reviewing best practices.
- Assessing risk based on the severity and likelihood of credible scenarios.

PSKM supports effective hazard identification and significantly impacts the quality of HIRA. Conducting a quality HIRA involves:

- Compiling accurate and complete process safety information.
- Ensuring review team members are knowledgeable in the process being analyzed and in the PHA process.

Several examples of PSKM impacting the second RBPS Pillar - Understand Hazards and Risk – are presented in Table 3-3.

Table 3-3 Examples of PSKM Interactions on the Second Pillar of RBPS

RBPS Element	Example PSKM Interactions
Process Knowledge Management	• Understanding of the change management system to ensure that updates to the organization's P&IDs are timely and complete. • Understanding of the Hazard Identification and composition information contained in Safety Data Sheets. • Knowledge of chemical inventories and storage. • Understanding of critical safety systems, their design basis, and operational parameters. • Development of operational expertise. • Process knowledge to develop personnel competencies across the organization.
Hazard Identification and Risk Analysis	• Training that includes: • Thorough understanding of the organization's risk management system for team members. • Understanding of HIRA activities into project and process life cycles. • Understanding the hazards of the process they are responsible for operating/supporting/designing (the outcome of the HIRA).

PSKM and HIRA are linked throughout the Facility Lifecycle. In the Project Development Stage, most companies will not undertake a HIRA but instead they will identify significant hazards and risks. Most of the PSKM activities in this stage are mainly related to the Capture and Organize steps of PSKM as described earlier in the chapter. HIRA is typically performed in the Design Stage based on information, documentation and knowledge gained up to that point in the Process Lifecycle. As changes are introduced to the project, PSKM continually supports successful HIRA activities.

Data collection and knowledge development in the initial stage enables early hazard identification. This process continues throughout the design and construction phases resulting in the capture of the foundational Process Safety Knowledge. In the Operational phase, continuous observations along with changes or advancements in process technology/equipment require additional

HIRA activities that are supported by the change management process and the competency of the personnel involved with the change. Maintaining a robust PSKM system is instrumental to achieving a safe design in the design phase and subsequent hazard studies including decommissioning of either a facility or a process. Finally, knowledge of special or residual hazards becomes critical to safe decommissioning of process units or safe change over to different technologies.

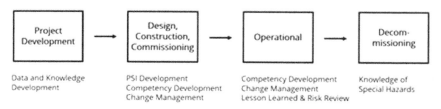

Figure 3-4 Integration of PSKM Into HIRA Activities

3.5   Manage Risk

The third pillar under RBPS is Manage Risk. PSKM has a strong tie to managing risk by developing and sharing knowledge through standards, practices, and competencies. This pillar covers:

1. Operating Procedures
2. Safe Work Practices
3. Asset Integrity and Reliability
4. Contractor Management
5. Training and Performance Assurance
6. Management of Change
7. Operational Readiness
8. Conduct of Operations
9. Emergency Management

PSKM is critical to successful implementation of three RBPS elements under the third pillar: Operating Procedures, Training & Performance Assurance, and Management of Change (MOC). Operating Procedures are key in dissemination of Process Safety Knowledge in each phase of the operation. Similarly, Operating

Procedures provide written instructions forming a collection of information that can be turned into Process Safety Knowledge using a robust PSKM System. The PSKM System enables Training and Performance Assurance by setting the level of expertise requirements and measuring their completion. Without a robust PSKM System, the MOC process may not be able to deliver updated Process Safety Knowledge as changes are made.

3.5.1   Operating Procedures

Operating Procedures and PSKM are closely linked. Operating Procedures are written instructions that (1) list the steps for a given task and (2) describe the way the steps are to be performed [23]. PSKM helps ensure:

- Accurate and complete written instructions for the manufacturing and/or operating processes
- Explanation of the safe operating limits and potential consequences of deviation from those limits
- Chemical safety and health considerations
- Description of key process controls and safeguards
- Steps to take to prevent releases of hazardous chemicals and other forms of energy

The written instructions in the Operating Procedures must be embedded in the organizational knowledge and knowledge of the organization's employees who operate and maintain the process. Effective PSKM can be used to improve Operating Procedures as presented in Table 3-4.

Table 3-4 PSKM Impact on Operating Procedures

PSKM Source	Impact on Operating Procedures
Consequence information described in PHA high risk scenarios.	• Provides critical safety information within the operating instructions in procedure steps or precautionary statements where potentially hazardous conditions exist. • Enables the administrative control of hazardous chemical inventory levels.
Mitigation safeguards included in PHA high risk scenarios.	• Provides information related to process limits. • Gives consequences of deviations from these limits.
Alarms	• Emphasizes process and safety criticality for temperatures, pressures, flows, and other operational parameters. • Helps guide the operator to safely run the process.

Safety Data Sheets	• Contains vital information including:
	  o Chemical hazards.
	  o Precautions necessary to prevent exposure.
	  o Control measures to be taken if physical contact or airborne exposure occurs.
	  o Quality control for raw materials.
	  o Special or unique hazards.
	• Provides essential information for inclusion in safety and health considerations of operating procedures as well as for use by PHA teams, process engineers, trainers, and incident investigation teams

The review of the PSK contained in operating procedures can be completed concurrently with operating procedure reviews, process technology information reviews and/or periodic PHA revalidation. This review process will assist the reviewer in identifying any gaps between the Operating Procedures and the Process Safety Knowledge that require updating. Chapter 4 presents an example of a Process Technology Information Review Guide and an example of a chemical hazard information tracking measurement tool.

A list of critical interlocks might be provided in an operating procedure. However, how the critical interlock activates and prevents a potential consequence of deviation is part of the Process Safety Knowledge. A PSKM system provides an infrastructure to capture this knowledge in the Process Knowledge Management element, organize it in the Hazard Identification and Risk Analysis element and delivers it through a brief summary in the Operating Procedure element along with the Training and Performance Assurance elements.

3.5.2   Training and Performance Assurance

The objective of training is to enable employees to meet some minimum initial performance standards, to maintain their proficiency, or to qualify them for promotion to a more demanding position [23] PSKM supports this element by providing a Process Safety Knowledge structure along with the competency requirements based on the commitment to process safety. An effective PSKM System enables organizations to learn and disseminate knowledge continuously. Through performance assurance, employees demonstrate that they have understood the training and can apply it in practical situations. Again, a successful PSKM System supports performance assurance by defining roles and responsibilities and validating the program effectiveness.

3.5.3   Management of Change

Changes that are not replacements-in-kind require a formal review assessment and approval process. An effective PSKM system enables the organization to assess the need and level of risk assessment. The review process should include:

- Assessing the impact of the change on existing safety systems and process equipment.
- Updates needed to procedures, drawings, process controls and other process knowledge.

It is very important to promptly update process safety information and process knowledge when a change is made. Process changes must be incorporated into the organizational knowledge and knowledge of employees who operate and maintain the process. This cycles back to the competency element discussed in Section 3.1. Without a clear and concise understanding of the Process Safety Knowledge element and its importance, changes cannot be effectively managed.

To keep the Process Safety Knowledge updated and accurate, it is critically important to identify all the documents impacted by a process change. To help the Management of Change (MOC owner identify all the documents to be updated, a "Document Change Checklist" can be included as part of the MOC management system. The checklist can be used to identify, for example, procedures, forms, drawings, training, mechanical and electrical files to be updated because of the change. Chapter 4 presents a checklist example. Once the documents that require updating are identified, an action plan should be put in place with a due date and person responsible. These action plans should be tracked for timely completion and closure verified. The PSKM system extends beyond keeping PSI updated and accurate. An effective implementation of PSKM helps with the dissemination of knowledge after the changes occur and as new knowledge is attained. PSKM supports MOC by providing a framework for updating employees knowledge after the change occurs. This augmentation of knowledge can occur not only through new experiences but also simulations and related risk assessments.

In some facilities, even though an MOC system exists, it might be regarded as a documentation tool for regulatory purposes. This view of MOC systems may be due to lack of understanding, lack of competency, and/or a weak process safety culture. Having an effective PSKM system will improve and enhance the MOC process. When a change occurs, new knowledge needs to be captured, organized, and provided to all employees impacted by the change. As the organization's maturity with the PSKM System improves, there is a corresponding improvement in the execution of MOCs. A PSKM System can also help with establishing an effective process to build competency to help anticipate effects of changes and communicate that to employees.

A strong Process Safety Culture, in which the MOC system is trusted and valued will eliminate re-work and re-verification of system information at critical operational demand periods. When the MOC system is fully utilized, it ensures that the PSKM is continuously improved, updated, and managed as changes are fully evaluated for their impact to safe operations. Full utilization of the MOC system comes through an organization understanding the value in the MOC system and fully incorporating it into their Process Safety Culture. The MOC process should include a formal risk evaluation for every proposed change to determine the level of safety review required because of the change. The level of safety review could range from a minor safety review to a full HIRA study depending on the assessment depth required. A successful PSKM System implementation would enable these decisions and provide the appropriate level of safety review while assessing the impact of the change.

For example, a proposed change might require a full PHA study if the change:

- Extends PSM boundaries.
- Potentially increases personnel exposure and other forms of damage.
- Increases chemical inventory.
- Changes process limits and consequences of deviation.
- Affects known safeguards such as a change in relief valve requirements or a change in materials of construction.

A well-implemented PSKM system would improve the MOC risk assessment process by bringing together the right people, with the right competency based on the captured and organized knowledge.

While managing risks, various action items and recommendations are generated. PSKM plays an important role in closing identified gaps. Having an effective PSKM will allow a more thorough review of the change through a full understanding of the process and how the proposed change can impact the safe operation. In addition, a strong PSKM culture means that the MOC program will promote continuous updating of PSKM including updating PSI, training, and understanding of how the process is changed. This culture will enable people in the organization to remain knowledgeable and ready to evaluate a new change when it is proposed in the future. The PSKM System and MOC program are closely linked. An organization cannot have an effective MOC program without the knowledge / wisdom that a PSKM System provides. Simultaneously, a strong MOC program will keep that PSKM System fresh and evergreen through gap identification, action assignment, and documentation of actions related with the change.

Companies with effective PSKM can improve the following:

- Efficient tracking of action items from generation to closure with verification that completed actions function as intended.
- Consultation with appropriate personnel to determine the best means of closing out identified actions.
- Recordkeeping of closure methodology and acceptance, to be able to clearly demonstrate action item implementation.
- Training of personnel on the change and the potential for change to PSK.

3.6   Learn from Experience

The fourth and last pillar under RBPS is Learn from Experience. Learning from experience and PSKM are linked with the goal of sharing experiences to gain knowledge. This pillar covers:

1. Incident Investigation
2. Measurement and Metrics
3. Auditing
4. Management Review and Continuous Improvement

All elements of the fourth and last pillar of RBPS will be discussed in this section because of their critical tie to keeping PSKM current and evergreen. The elements within Learn from Experience include applying lessons learned from past incidents, monitoring/ assessing the health/effectiveness of the site's overall RBPS implementation and identifying opportunities for improving PSKM to bridge gaps that have been identified in the RBPS.

3.6.1   Incident Investigation

Effective PSKM allows for the efficient use of incident investigation information and the identification of linked common causes. Sharing incident investigation outcomes/findings/actions across all facilities that a company operates is another added value of a successful PSKM System. Outcomes/findings/actions from an incident that happened at one site can be used to prevent the same or a similar incident from occurring at another site. In fact, effectively sharing not only knowledge but high-quality knowledge across facilities requires a mature PSKM System in the organization. Knowledge can be and in fact is often shared through all kinds of communications, notices, and other means. It is the effectiveness of that sharing that a mature PSKM improves. Subsequently, a successful implementation of PSKM enables clear identification of gaps in current safety systems and the development of risk mitigations. Similarly, incident investigations are going to identify gaps in the current PSKM that allowed for a lack of knowledge/understanding of the process. As an outcome of an effective incident investigation, actions for improvement/gap closure will be identified. Those recommended actions will close the gaps in PSKM and strengthen the overall PSM system. In fact, the strong connection between Incident Investigation and PSKM can include other elements such as updating operating procedures. The updated procedures capture the knowledge gained from the incident investigations. When employees are trained in the updated procedures, this new knowledge is passed to them. Chapter 5 presents a discussion regarding the health of the PSKM System. A healthy PSKM will support successful incident investigations.

Incident investigations that follow a cause and consequence pair identify gaps and improvement opportunities after that gap led to an incident. In contrast, near miss investigations allow an organization to prevent a potential future incident. In fact, high-hazard industries such as nuclear power and aviation, are particularly sensitive to accident precursors [2].

Accident precursors are defined as events that must occur for an accident to happen in each scenario but have not resulted in an accident so far. Tracking these early warning events through the Measurement and Metrics as well as Continuous Improvement elements are discussed Section 3.6.2. Incidents trigger reactive actions but near misses can allow proactive improvement of a site's PSKM system. Not all companies put enough effort into nor see the value in thorough near miss investigations. However, emphasizing near miss identification and investigation may prevent an incident from occurring because PSKM and RBPS will be much more robust based on the knowledge captured in the process. Incident and near miss Investigations promote implementation, maintenance, and monitoring of an effective PSKM System in the organization by:

- Identifying gaps and discrepancies in the process safety information and knowledge.
- Identifying the common root causes and causal factors of incidents and near misses.
- Sharing of lessons learned from incidents and near miss investigations across the organization to develop a knowledgeable and process safety conscious workforce either within a facility or across an organization.
- Learning from the incidents at other facilities in the industry, from incidents in other industries, or from different publications such as CCPS

Process Safety Beacon [24], CSB reports [25], EPSC Learning Sheets [26], and other sources.

3.6.2   Measurement and Metrics

Organizations should establish leading and lagging Key Performance Indicators (KPIs) to monitor the health of the overall PSKM system and proactively identify potential issues early. Metrics provide information about issues that can be corrected or improved. The need for organizations to establish KPIs is universally accepted but the connection between KPIs and PSKM may not be clearly defined. Organizations should establish KPIs because they will identify gaps/opportunities for improvement prior to an incident occurring. This gap identification process allows for the PSKM to be updated or better utilized throughout the organization so process incidents are prevented. KPIs tell an organization how effective their PSKM is at supporting their RBPS program. Examples of KPIs in PSKM are:

- Incidents or near misses that identify common root causes or that involve failures of like equipment within the facility or across the organization.
- Incidents that involve failures of or affect safety critical equipment.
- Development and Sharing of Lessons Learned Database.
- Competency development to expand pool of experts for PHA team participation.
- Evaluations of findings published by safety organizations that could have similar bearing on the organization.
- PHA recommendations.
- Open MOCs.
- Open safety actions.

A detailed discussion of KPIs and their implementation as part of the PSKM system will be provided in Chapter 5.

3.6.3 Auditing

Once a PSKM system is established, this system should be audited to ensure it remains effective. Auditing of a PSKM system has historically focused on the documentation of Process Safety Information only. However, to adequately audit PSKM processes, organizations must also audit the level of understanding within the key personnel who interact with and support a process. It is important to assess whether the information and knowledge is adequate to meet the minimum recommendations of the RBPS elements at each point of the unit's lifecycle. The goal of such an audit is to observe the Process Safety Information in action and see the knowledge reflected in operations, training programs, MOCs, and incident reviews. Chapter 5 presents audit criteria for an effective PSKM process.

3.6.4 Management Review and Continuous Improvement

Effective PSKM also includes Management Reviews and Continuous Improvement. Management reviews involve a management team (engineering managers, operations managers, safety managers, plant managers, and other managers such as maintenance managers) reviewing KPIs, audits, investigation results and then prioritizing improvement actions and capital funds as needed to close the gaps and strengthen PSKM throughout the organization. Management reviews promote an implementation, maintenance, and monitoring of an effective PSKM System by:

- Striving to continuously improve by understanding actual performance versus target goals.
- Conducting field inspections which provide information about any existing or potential hazards, provide an understanding of employee concerns, and help management gain further understanding of job requirements and the workplace environment.

These activities interact with the first Pillar, Commit to Process Safety. A strong leadership indicates commitment to process safety. Commitment to process safety is demonstrated through regular (ideally occurring monthly or quarterly) management reviews that identify and prioritize areas for continuous improvement and track resolution of associated action items. Once the management reviews the program on a regular basis, documents it as a learning opportunity and prepares a plan to improve it, the loop for the RBPS will be complete. This last step will indeed promote commitment to process safety and strengthen the process safety culture of the whole organization. The PSKM

System will support all these critical activities through the combination of knowledgeable people and well-developed processes to bring about safe operations.

3.7   Chapter Summary

In Summary, there is a connectivity between Process Safety Knowledge Management and the four pillars of RBPS. Although four key elements (Process Knowledge Management, Operating Procedures, Training, and Management of Change) are discussed in detail, all RBPS elements interact with PSKM. The reason for this strong interaction lies in the framework of RBPS. Through the effective use of four pillars, multiple elements of RBPS support each other and enable the full RBPS framework. Knowledge is generated in the Understand Hazards and Risk pillar which includes Process Knowledge Management, which is a part of PSKM but not the only one. Knowledge is subsequently shared and provided in the Manage Risks pillar. Knowledge is applied through the Commit to Process Safety pillar. And finally, knowledge is updated through the Learn from Experience pillar.

Sustaining Process Safety Knowledge should be fully integrated into everyday activities. This knowledge ownership should be assigned to the groups that derive the most value from the information. Knowledge owners understand the limits of the data and can share that information across the facility or organization.

3.8   Introduction to the Next Chapter

In the next chapter, key steps in developing a Process Safety Knowledge Management (PSKM) program and potential obstacles to overcome during its implementation are introduced. Effective PSKM Systems make necessary knowledge available to everyone who needs it, when they need it, and at the right level of detail. Key information will also include:

- Measurement tool to track PHA action closure status.
- Chemical hazard information guide for PSKM in Operating Procedures.
- Process Technology Review Guide and Hazard Registries.
- Discussion of setting up an effective PSKM system to support successful management of change.
- An example PSKM RACI chart.

4 Developing and Implementing PSKM

"Knowledge is of no value unless you put it into practice." Anton Chekhov

4.1 Introduction

In Chapter 3, effective PSKM is defined as a system for capturing, organizing, maintaining, and providing the right Process Safety Knowledge (PSK) to everyone who needs it, when they need it, and at the right level of detail (Figure 3-1).

A successful PSKM System is comprised of four elements: Capture, Organize, Maintain and Provide. This chapter provides a detailed discussion of the capturing, organizing, and providing steps (three of the key PSKM System elements) and examples in developing a PSKM program and potential obstacles to overcome during PSK implementation. Maintaining Process Safety Knowledge will be discussed in Chapter 5.

PSKM begins with the data and information, commonly referred to as the PSI, related to the process that fully describes the Chemical Hazards, the Process Technology including chemical reactions and unit operations, and the Process Equipment. PSI transitions to PSK when it is used to educate and inform personnel throughout the organization. Knowledge coupled with practical experience, as the workforce operates, engineers, and manages the process, ultimately results in a deep understanding of the process leading to operational wisdom resulting in safer and more efficient operations.

An effective PSKM System is essentially a tool that makes necessary Process Safety Knowledge available to everyone who needs it, when they need it, and at the right level of detail. Understanding and managing risk depends upon accurate Process Safety Knowledge. It is not enough to have an archive of documents with titles which match requirements found in process safety management regulations and standards around the world (such as Seveso Directives in Europe, OSHA regulations in the USA, Official Mexican Standards (NOMs) in Mexico, and CSA Standards in Canada). A successful PSKM System embodies transformation of data and information into knowledge and ultimately to operational wisdom allowing each person to apply it in their individual job responsibilities and tasks. Knowledge must be developed and presented in a format which satisfies the needs of the final users. It is critical that sufficient competent resources are provided to develop Process Safety Knowledge, support the PSKM System, and ensure the knowledge remains usable.

One of the most challenging areas in PSKM is ensuring the knowledge and wisdom developed within people is not lost when people change their position within the company or leave the organization. Addressing this challenge successfully requires planning and foresight as well as proper systems and people to support them.

Systems must be in place to assure people who have the required skills, educational backgrounds and/or professional experiences succeed as recognized resources. PSKM should be introduced to new employees in onboarding programs and incorporated into professional development plans for all employees.

Onboarding programs that incorporate Process Safety Knowledge Management equip people with the resources to succeed in their jobs. Structured walk-throughs of the PSKM System and access to information increase productivity and safety performance and ensure people quickly find information they need to successfully perform their jobs.

There are three ways an organization benefits when people use Process Safety Knowledge Management Tools:

1. The location for finding information will become a knowledge tool.
2. People who use knowledge management tools will support others in doing the same.
3. The process will promote continual improvement if it contains provisions for feedback and refreshing of the knowledge.

The PSKM tools become the single source of truth rather than "picking up" PSK via informal means.

There is no automatic way to "upload" or "transition" knowledge and wisdom from one person to another. Structural processes must be intentionally created for experienced personnel to "pass on" their knowledge and wisdom. Roles and responsibilities must be established for creating a systematic knowledge management system. In most cases, the organization's managers have the overall responsibility to identify the subject matter experts who are critical to performance and recruit or groom individuals who will manage that knowledge. To achieve these goals, human resources (HR) management personnel help the organization recruit talented individuals to fill these roles.

Structural processes to "upload" or "transfer" information include:

1. Clearly defining job roles, responsibilities, and tasks which translates to expectations for personnel
2. Determining the skills, knowledge, experience needed at each job level (entry, advanced, senior, specialist, and others) which is essentially what amount of PSKM is expected in each level of the organization
3. Developing detailed training materials (standard work) and engaging training programs for each job level to teach people about the systems and information available to them, how they can utilize it, and their expectations
4. Requiring development plans for employees to gain skills, knowledge, experience to progress through the PSKM transfer
5. Encouraging or requiring mentor relationships to speed transfer of PSK (senior employees mentor entry level employees in the same or similar job function)
6. Providing opportunities to cross train where PSK could be obtained or applied in a different way (this might be a part of their development plan)
7. Supporting both internal and external opportunities to gain knowledge and experience (conferences, continuing education, seminars, certification programs, job shadowing)

Organizations need a proactive method for purposefully developing employees and transferring knowledge. A structured program for PSKM is essential to provide new employees with an accelerated learning curve so that they can effectively learn the lessons of past incidents at the onset of their careers and maintain this knowledge over the duration of their careers.

The details of this section to develop and implement PSKM are divided into different sections such as establishing people, building systems to capture, organize, and provide PSK, and finally combining organizational structure with the process safety culture.

4.2 Resources for Capturing, Organizing and Providing PSK

Resources are needed to capture, organize, and provide the PSK. These required resources are composed of the right people within the organization to build and guide the knowledge development, together with a defined process, and an overall plan.

Having clearly defined roles and responsibilities enables management to identify the types of qualified people needed to successfully implement the PSKM system. Examples of qualifications include certifications, industry and company experience, skills, and knowledge.

Senior management commitment is essential to successfully implement a well-functioning PSKM System. Once management defines its goals and selects the PSKM Project Manager, the next step is to appoint a sponsor at the senior management level who will provide guidance and support to the PSKM implementation team.

4.2.1 PSKM Implementation Teams

An implementation team for PSKM may be required for a large project. Another option is to assign specific site personnel whose job roles and daily responsibilities are to capture and provide information and knowledge. If so, then senior management is responsible for defining which team members are responsible for the different aspects of PSKM as well as providing the team with the necessary training and resources to meet their expectations. For example, a site might require plant engineers to be responsible for providing guidance on original equipment manufacturer's manuals (OEM) and equipment specifications that need to be placed into the process knowledge. MOC coordinators or EHS professionals may be responsible for identifying changes which impact process knowledge as well as assigning personnel responsible for providing that information. If the site has a MOC program and a PSKM program defined, then there may be no need to assemble a team. However, if the PSKM program is not defined or established, a formal team may need to be assembled.

The PSKM Implementation Team members could consist of personnel in the following roles:

- Researchers and Scientists
- Process and Control / Instrumentation Engineers
- Operations Personnel
- Maintenance and/or Asset Integrity Personnel
- Mechanical Designers who built the plant
- Process Safety Engineers
- Occupational Health and Safety Personnel
- Resident Contractors
- Information Technology Personnel
- Internal or external PSKM consultants and/or implementation specialists

The team must be appropriately sized to meet its goals and be effective. Industry guidance recommends these three points to consider when putting together an effective team:

1. Types of tasks the team will undertake to define what type of skills are needed
2. Team composition to ensure individual members are appropriate to the task
3. Team size

While research on the optimal number of people serving on a team is not conclusive, a team of five or six is optimal within organizations for task coordination. However, having a good team depends on more than optimal team size.

Several examples for PSKM Roles and Responsibilities for PSKM implementation, combined from different guidance, are presented in Table 4-1. The Chief Knowledge Officer is the highest position within PSKM. Communities of Practice (COPs) are organized groups of individuals who share an interest in a defined area and want to coordinate efforts to achieve specific goals. COPs focus on sharing best practices and creating new knowledge to advance the domain of professional practice.

The Project Management Institute's Guide to the Project Management Body of Knowledge (*PMBOK® Guide*) helps companies to standardize and document practices that consistently work [27].

Table 4-1 Example PSKM Roles and Responsibilities

Role	Responsibilities
Chief Knowledge Officer	Accountable for the overall PSKM strategy, planning, and implementation
PSKM Project Manager	Executive who manages the implementation of the PSKM initiatives
PSKM Champions	• Promote PSKM in the workplace • Facilitate Communities of Practice
PSKM Navigators	• Know where PSKM is located • Connect people who need knowledge with systems and people who have knowledge
PSKM Stewards	• Responsible for ensuring PSKM updates are made following Management of Change • Track changes for follow-up and validation
PSKM Editors	Manage format and language of knowledge so users can easily use it

4.2.2 RACI Charts

A responsibility matrix, or RACI chart, can be generated once the PSKM roles have been defined. The RACI chart will define whether people involved in the development and implementation of the PSKM System will be Responsible, Accountable, Consulted, or Informed (RASI) for the required tasks and decisions. RACI chart roles and responsibilities are defined in Table 4-2. Table 4-3 presents an example RACI chart for PSKM implementation.

Table 4-2 RACI Role Definitions

Role	Definition
(R) Responsible	The person or people responsible for performing the task, that is, the actual person doing the work to complete the task. A task must have at least one responsible team member, who oversees decision-making. A task can have more than one responsible team member.
(A) Accountable	The person who is ultimately accountable for the task being done in a satisfactory manner and who must sign-off the work the Responsible person produces. A task can have multiple responsible parties; however, it is best to have only one who is accountable for individual tasks to avoid confusion and allow faster-decision making.
(C) Consulted	Those people whose input is used to complete the task, thus communication with this group will be 2-way in nature.
(I) Informed	Those people who are informed as to the status of the task, thus communication with this group is 1-way in nature.

Table 4-3 Example PSKM RACI Chart

Role / Job Title	Mentoring	Leads PSKM Project	Knowledge			
			Capturing	Obtaining	Maintaining	Providing
Chief Knowledge Officer (CKO)	A	A	A	A	A	A
PSKM Project Manager	C	R	C	C	C	C
PSKM Champions	R	I	C	C	C	C
PSKM Navigators	C	I	C	C	C	R
PSKM Stewards	C	I	R	C	R	C
PSKM Editors	C	I	C	R	C	C

(R) Responsible; (A) Accountable; (C) Consulted; (I) Informed

4.2.3   Knowledge Resources for Development and Implementation

Knowledge resources available to organizational teams to capture, organize and provide the PSK are noted in Figure 4-1. Example resources for PSKM Maintenance are provided in Chapter 5.

Capturing Knowledge	Organizing Knowledge	Providing Knowledge
• Process and Mechanical Design Specifications • Piping and Instrumentation Diagrams • Hazard Risk Analyses • Consequences of Deviation Tables • RAGAGEP • Original Equipment Manufacturer Manuals • Relief System Design and Design Basis • Material and Energy Balances	• Company Intranets • Skills Directories • Knowledge Repositories • Hazards Register	• Operating Procedures • Standard Operating Conditions and Limits • Training • Chemical Properties and Hazards • Safety Data Sheets • Hazards Register

Figure 4-1 Example Resources for PSKM Development and Implementation

4.2.4   Logic Models for Developing and Implementing PSKM

A logic model is a tool that can be used to develop and implement the PSKM System [3]. Logic models are graphic illustrations of the PSKM Implementation Plan and show the relationship between the planned work and anticipated results. Figure 4-2 presents a logic model.

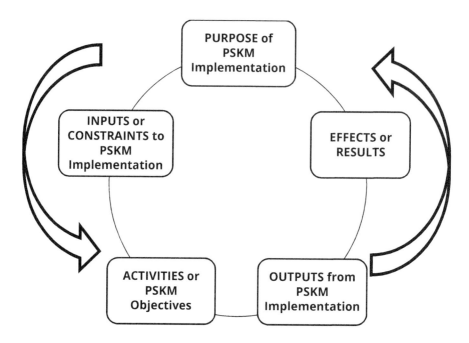

Figure 4-2 Sequence of Actions of PSKM Development Process

- The **Purpose** is a broad statement of the PSKM strategy and objectives.
- **Inputs** are the resources needed to accomplish the PSKM implementation or support the project. Examples are people, technical information, software, elements of Capturing knowledge explained in Section 4.1.5.
- **Constraints** are barriers to PSKM objectives such as capital, time, resources.
- **Activities** are the action components or process objectives such as defining how the PSKM will be governed including roles and responsibilities, plans and schedules, process flow charts.
- **Outputs** are the direct evidence of having performed activities or the intended accomplishments of PSKM implementation.
- **Effects** are the Results, Outcomes, Impacts of the implementation, the intended outcomes.

4.2.5 Required Input to Build the PSKM System

Figure 4-3 shows typical Inputs to a PSKM logic model for the distinct types of information that need to be captured:

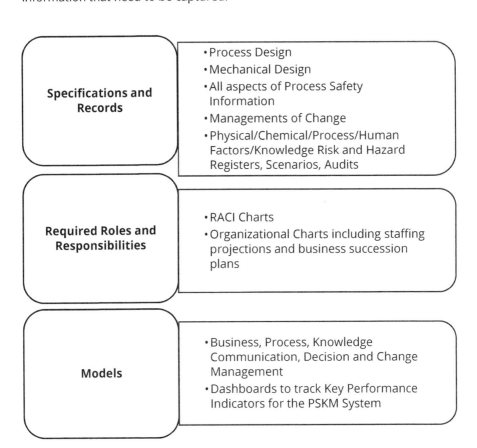

Figure 4-3 Inputs to a PSKM Logic Model

Activities related to a logic model for building a PSKM System are defined in Table 4-4.

Table 4-4 Example Activities for Building PSKM Management System

Activities/Resources	Definition
Change Management Modeling	Activity of representing how processes of an enterprise undergo change.
Dashboards	Graphic summary of various pieces of vital information, typically used to give an overview of a business or group.
Hazard Compilation	Inherent chemical or physical characteristic that has the potential for causing damage to people, property, or the environment.
Knowledge Communication Modeling	Activity of representing information, abilities, and ideas as shared across different areas of a business.
MOC – Management of Change	A management system to identify, review, and approve all modifications to equipment, procedures, raw materials, and processing conditions, other than replacement in kind, prior to implementation to help ensure that changes to processes are properly analyzed (for example, for potential adverse impacts), documented, and communicated to affected personnel.
RACI Chart Development	Graphical means to identify a team's roles and responsibilities for any task, milestone, or project deliverable.
Risk Register Development	Document used as a risk management tool and to fulfill regulatory compliance, acting as a repository for all risks identified and contains additional information about each risk, e.g., nature of the risk, reference, owner, and mitigation measures.
Workflows	Systems for managing repetitive processes and tasks which occur in a particular order.

4.3   Capturing Knowledge and Information

The types of knowledge and information to capture are shown in Figure 4-4.

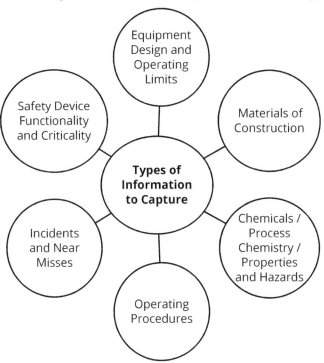

Figure 4-4 Types of Information to Capture

The sources used to capture information are shown in Figure 4-5:

Figure 4-5 Information Sources

Captured information can be stored in data repositories. Maintaining and reviewing data repositories will be discussed in Chapter 5.

Tacit knowledge can also be captured from within an organization. Capture processes could contain customizable information entry forms about a project with unique identifiers or best practice documents categorized by subject. Once entered, knowledge from previous projects can be accessed by employees and applied to new projects. Best Practices can be used to capture tacit knowledge in an organization:

- Promoting a knowledge sharing culture
- Incentivizing knowledge creation and sharing
- Automating knowledge creation with intelligent applications
- Creating formal and informal groups to discuss work problems

Table 4-5 presents an example of a logic model for capturing knowledge for building a PSKM system.

Table 4-5 Example Logic Model Components to Capture the Knowledge

CAPTURE THE KNOWLEDGE				
INPUT SOURCE	**CONSTRAINTS**	**ACTIVITIES**	**OUTPUTS**	**EFFECTS**
Information Type: CHEMICALS / PROCESS CHEMISTRY / PROPERTIES AND HAZARDS				
Technical Staffing	Budgets	Compilation of Chemical Properties Data from Technical Reports, Safety Data Sheets	Effects of inadvertent mixing of two chemicals evaluated.	Inadvertent mixing hazards documented.
Information Type: EQUIPMENT DESIGN AND OPERATING LIMITS				
Mechanical Integrity personnel	Time Commitments	Compile Mechanical Design Information	Description of location and operation of controls, maintenance schedules, parts numbers, specifications extracted	Controls, preventive maintenance schedules and parts numbers documented.
Information Type: INCIDENTS AND NEAR MISSES				
Incident Reports	Resources	Collect project or process risk, history	History and resolution information collected	Hazard and Risk Registry Documented

A success story in capturing knowledge and information is described in Success Story 4.1.

Success Story 4.1 – Siemens AG

Siemens AG – Siemens found it needed to share knowledge more effectively with its nearly half million employees around the world [28]. Its success story in knowledge management began with the formation of Communities of Practice which were composed of individuals responsible for creating, refining, and maintaining knowledge in their respective areas of expertise. Siemens utilized a database repository, search engine, and chat room. Online entry forms were provided for employees to enter knowledge they felt was useful into the repository. Employees could search the database and contact the author for further details. Incentives were given to employees who provided knowledge into the system. The project was completely supported by top management. Siemens found the database to be a valuable knowledge repository available to all employees. It enhanced global collaboration. Siemens' success factors include:

- Defined business needs.
- Managers and users engaged in its development.
- Training was carefully designed.
- There was a reward structure for knowledge contributions.
- Financial benefits.
- Top management support.

4.4   Organizing Knowledge and Information

Information must be organized to know what knowledge a company has for the purpose of making it accessible to others. The purpose of knowledge organization is to ensure effective retrieval, use and management of the organization's information.

To accomplish this purpose, organizations require personnel with specific skills to collect, organize, and update the knowledge. Knowledge organization systems serve to connect people to knowledge. People must know where they can find the knowledge important to them without having to wade through unrelated material.

Knowledge can be organized using a model known as "LATCH", an acronym for Location, Alphabetical, Time, Category, Hierarchy, given that the LATCH is implemented within modern data management, maintenance, and presentation systems. Table 4-6 presents these five methods [29]. Organizations must decide which method or methods would work best for their PSKM System development. Some organizations may choose multiple methods for diverse types of information or use subsets of one method within another. For example, spare parts stored within a maintenance warehouse might be organized by type of equipment (category), sorted by order (alphabetical), and organized by place (location). An Incident Investigation Registry might be organized by plant (location), incident date (time), and incident classification (hierarchy).

Table 4-6 Organizational Methods

Organization Method	Description	Use Examples	Advantages	Disadvantages
Location	Visual diagram or map of the information.	Emergency Exit Routes Emergency Response Equipment Location Location of parts within a warehouse	Assists with orientation and providing directions.	Requires visualization of the area which may be a constraint to facility visitors or contractors.
Alphabetical or Numerical	Information is sorted by name, department, subject, title.	Safety Data Sheets Equipment Tags and numerical Identification number	Simple, easy, and effective Useful for sorting substantial amounts of information	Requires knowing the name or topic being researched - Terms must make sense to the searcher
Time	Information is arranged in the order in which it occurs.	Incident Databases DCS Data Incident Investigation Timelines Understanding Incident investigation sequence of events	Useful for finding information in a chronological pattern Useful for describing a chemical production process	May present challenges when the user is unsure of the chronological history

Table 4-6 Organizational Methods - Continued

Organization Method	Description	Use Examples	Advantages	Disadvantages
Category	Information that is sorted by similarity.	Process Hazard Analysis Reports Equipment tag descriptors	Grouped by similar importance.	Users must be aware of the Category names and the types of information stored in each category
Hierarchy	Information organized by importance or rank.	RAGAGEP National Local and Company Codes that govern equipment design Incident Command Hierarchy within an organization	Shows who reports to whom Information is shown in order of magnitude	May discourage collaboration and information sharing.

When organizing PSK, the most important thing is to make sure the knowledge is organized in a way users will be able to quickly find and access it. The organizational method should be from the perspective of the users and not necessarily the subject matter experts. For example, if the organization expects the knowledge to be used by engineers and operators, it should be based on their thought processes and how they approach information gathering which is not necessarily the same way EHS personnel would think about the same information.

Table 4-7 presents example inputs and activities to organize the knowledge. Inputs to Table 4-7 are outputs from Table 4-5. Table 4-7 outputs may require learning and development of internal expertise within an organization. These outputs should also be reviewed for alignment with the overall PSKM goals.

People will know about information if they have access to it. People will develop knowledge as they access the information.

Table 4-7 Example Logic Model Components to Organize the Knowledge

		ORGANIZE THE KNOWLEDGE		
INPUT SOURCE	**CONSTRAINTS**	**ACTIVITIES**	**OUTPUTS**	**EFFECTS**
Information Type: CHEMICAL PROPERTIES / PROCESS CHEMISTRY				
Effects of inadvertent mixing of two chemicals evaluated.	Time Commitment	Identify fire and explosion hazards, reactivity hazards, toxicity hazards.	Chemical reaction risk matrix generated.	Hazardous Effects of Inadvertent Mixing documented for all chemicals present in the process.
Location and operation of controls, maintenance schedules, parts numbers, and equipment specifications.	Time Commitment	Identify software requirements for automated scheduling of preventive maintenance (PM), spare parts management, and training needs.	PM software obtained.	PM schedules developed, training materials for mechanics developed, spare parts planning developed.
Information Type: INCIDENTS AND NEAR MISSES				
History and resolution information collected.	Capital for Incident Investigation Registry Software	Identify software requirements for storing incident information and tracking action items	Central hub to organize registry put in place.	Registry information for prioritization, updates, tracking provided.

4.5   Providing Knowledge and Information

Providing knowledge and information can bring many benefits to an organization and can help to solve problems in an efficient way.

Knowledge can be provided in multiple ways such as through regular meetings to discuss lessons learned from a past event or situation, periodic sharing sessions, and/or access to real-time knowledge and expertise.

Providing access to the PSKM System should incorporate the ability to search using key words so the user can navigate from one folder to another and be able to search for specific information. The information in the PSKM System should be provided in a format using a simple logical structure that minimizes the time and effort to find the information needed.

Access to external knowledge systems can provide access to many RAGAGEP documents including NFPA Codes and Standards or API Recommended Practices.

Providing knowledge and information can also be built into a collection of questions and answers in an organization's PSKM system. Posted questions and answers to and from the organization's Subject Matter Experts can be captured in the PSKM System to preserve and organize the information so that it can be continually available. This practice will avoid repetitive questions. As applicable, question and answer topics can be incorporated into the organization's standards during their periodic revision processes.

Table 4-8 presents example inputs and activities to provide knowledge. Inputs to Table 4-8 are outputs from Table 4-7. Table 4-8 outputs may require learning and development of internal expertise within an organization. These outputs should also be reviewed for alignment with the overall PSKM goals.

Table 4-8 Example Logic Model Components to Provide Knowledge

		PROVIDE THE KNOWLEDGE		
INPUT SOURCE	**CONSTRAINTS**	**ACTIVITIES**	**OUTPUTS**	**EFFECTS**
Information Type: CHEMICALS / PROCESS CHEMISTRY / PROPERTIES AND HAZARDS				
Chemical contamination risk predictions generated	Information not adequately disseminated or maintained	Develop: • Training plans. • PPE Matrix • Chart categorizing hazardous effects of inadvertent mixing	Required PPE obtained Information communicated to Operations Staff, PHA teams, others	Chemical Mixture Compatibility Matrix provided PPE Matrix Developed
Information Type: EQUIPMENT DESIGN AND OPERATING LIMITS				
Preventive Maintenance software obtained	Time Commitment	Create computerized workflow solutions for preventive maintenance and spare parts management	Computerized maintenance management system populated	Use of computerized maintenance management system communicated, and training provided
Information Type: INCIDENTS AND NEAR MISSES				
Central hub to organize Incident Investigation Registry information implemented	Maintaining up to date information	Populate Registry information	Risks categorized, tracking system for resolution implemented	Use of the registry communicated, and training provided

Having identified information to be included, and after having decided upon a logical framework to provide the knowledge, it is also important to classify the resulting knowledge with respect to who is allowed to have access. In general, the classifications should include the following:

1. General Public Access
2. General Internal Access, but not General Public Access
3. Restricted (confidential) Internal Access, but not General Internal or Public Access
4. Tightly Restricted Internal Access, but not Restricted Internal, General Internal or Public Access

The issue with this type of classification is that it is possible for an access classification system to directly contradict the stated aim of providing the right information at the right time to the right person. Further, the time component should be considered and may lead to the possibility of temporary access at a higher-than-normal access level. This may apply to Contractors and to special Consultants such as incident investigators or legal counsel. However, as this aspect verges on security control of information rather than knowledge management and is partly a function of a business's special interests, knowledge access classification is not further considered in this book.

4.6   Organizational Structure to Build PSKM

A framework is presented for building the PSKM structure and the activities needed to transition from the current state to the final state. The structured process is composed of three states: Current, Transition, and Future. The Technology Requirement for software and tools, and Human Resources Support are also needed to transition to the future PSKM state. Figure 4-6 presents a structured process for PSKM development.

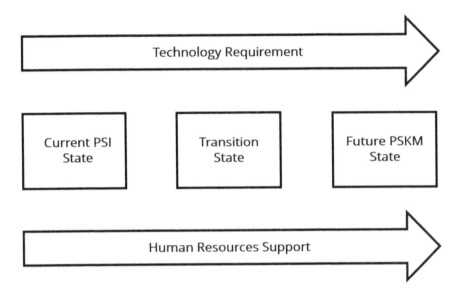

Figure 4-6 PSKM Organizational Structure

The steps necessary to establish a PSKM Organizational Structure steps are summarized as follows:

1. The Technology Requirement identifies systems, communication platforms, software, and other tools to be designed, developed/configured, and provided to reach the Future PSKM State. Most PSKM initiatives that rely heavily on software development fail, due to overwhelming complexity. Depending on the company's resources, one approach is to use a vendor platform as the starting point. And those platforms usually promote the idea of configuration, instead of coding.

2. The Current Process Knowledge State identifies how the process knowledge is currently managed.

3. The Transition State describes the plan to move from the Current Process Knowledge State to the Future PSKM State.

4. The Future PSKM State describes how the PSKM System will be managed once it is fully implemented.

5. Organizational Support involves people needed to reach and sustain Future PSKM State.

The organizational structure chosen for PSKM implementation could be either a flat or hierarchical structure. Both structures can work. In a flat structure, the decision-making process and communication can be faster, but productivity might be harder to track [30]. In a hierarchical structure, there are clear lines of communication but sharing of knowledge across lines might not be encouraged.

Whichever organization structure is chosen for the PSKM team, communication is key to prevent barriers in knowledge sharing.

The PSKM implementation team's work must be divided into specific action plans to determine, for example:

- Identification of the subject matter experts in process safety, engineering, and other technical and business disciplines
- Whether process knowledge and competencies are in place
- Who already has the knowledge or can provide it
- Whether there are any gaps in information

4.6.1   Steps to Implement PSKM

The process steps toward building a successful PSKM System involve planning, implementing, and improving the process. Table 4-9 describes a framework in eight steps that can be used to implement a PSKM system. This framework is consistent with, but defined in more detail than, the Plan-Do-Check-Act (PDCA) model [31].

Table 4-9 Steps to Implement PSKM

STEP	STEP DEFINITION	PLAN REQUIREMENTS
1	Establish PSKM Program Objectives	Identify and document input needed to meet desired goals.
2	Prepare for Change	To minimize any negative impacts from implementing PSKM, prepare for cultural changes to encourage knowledge sharing. This can be accomplished by recruiting PSKM Champions.
3	Define a High-Level Process	Understand how PSK will be identified, captured, validated, categorized, and provided.
4	Prioritize Technology Needs	Assess what kind of technology will improve and automate PSK activities. Understand what technology employees use today, what works and what does not work.
5	Assess Current State	Assess core components of PSKM: People, Processes, Technology, Structure and Culture to provide current state of PSK.
6	Build a PSKM Roadmap	Develop implementation milestones. Continue to review roadmap based on changing conditions and drivers.
7	Implement	Celebrate incremental advances, close gaps identified in assessment.
8	Measure and improve PSKM Program	Measure effectiveness by comparing actual results to target results. Establish baseline measurements prior to implementation and then compare results to see how performance has improved.

4.6.2   Planning for Business Changes for Successful PSKM Development and Implementation

Business plans list priorities, goals, milestones and expected results. Business models continually develop as companies react to changes to reduce risk or seize new opportunities. The ability to quickly react to change depends on the availability and retrievability of an organization's knowledge. Business cases for PSKM that connect PSKM goals to business goals, such as reduced risk, can be viewed as beneficial to an organization.

In addition to the investment of internal resources to build the PSKM, external resources will add costs that have to be considered, such as technology enhancements, expertise from specialized consultants, or additional headcount. An understanding of the additional capital or budgetary needs will help ensure the correct estimates have been made. Development of a PSKM business management plan can provide the model needed to understand the effect of implementation on the business. Figure 4-7 presents an example business management plan that has been compiled from several resources.

Required for PSKM	• PSKM Software or other methods to gather information and contact experts • Communi-cation Platforms • People
Business Goals	• Provide the right information to the right people at the right time • Improve Employee Training • Systems to Reach the Future PSK State
Expected Benefits	• Reduced costs associated with Process Hazard Analysis (PHA) • Improved plant reliability and reduced downtime • Sustain assets
Target Cost	• Investment cost dollars • Fit within company budget
Timeline to Implement	• Estimate length of time needed • Estimate hiring timeline based on organization structure transition
KPI Measures	• Identify indicators to be measured and frequency • Expected productivity measures

Figure 4-7 Example PSKM Business Management Plan

A business change management plan is a planning tool to help an organization prepare for external changes. The plan should contain updated goals, an understanding of the changes about to take place, and activities needed to meet new goals. An organization with dependable and adequate PSKM will have the flexibility to make quick business changes, make wise decisions, and operate safely when the change occurs.

The critical success factors significant for implementation of PSKM are listed as follows: [32]:

- Executive support
- Implementation and Training Plan for PSKM System Users
- Collaborative and knowledgeable employees
- Selection of knowledge management software or tools
- Learning culture
- Supply chain integration
- Comprehensive strategies
- Flexible organizations

4.6.3   Employee Mentoring and Retention

Employee retention (as a function of their ability to provide relevant knowledge and wisdom) is both critical for an effective PSKM System and good for all aspects of a business. Employees who stay with an organization will become knowledgeable and gain wisdom in PSK. Organizations will retain their developed PSKM experts.

Mentoring programs pair a mentor with a new employee to ensure new employees will learn from experienced employees. Mentees are more committed to the organization, which promotes better performance and employee retention [32].

There are many benefits that can be realized from a mentoring program focused on PSKM which include:

- Dedication to two-way communication programs with process safety experts
- Fostering risk awareness leading to a reduction of potential errors or mistakes
- Development of networking skills and a pooling of Process Safety Knowledge

4.6.4 PSKM Succession Planning

After PSKM is established, succession planning can follow. Most organizations have experts in process safety, engineering, and other technical and business disciplines. Organizations that conduct succession planning ensure successors will learn from the experts. With PSKM aligned with business goals, succession plans will result in the development of an effective network of PSKM experts. Organizations can accomplish this goal by first identifying their process safety experts and the candidates likely to become their successors. The process safety experts will then become mentors for the successors.

Examples of succession planning impacts on PSKM include:

- Assigning a PSKM Champion to lead implementation of succession planning
- Training the PSKM experts as mentors
- Providing job instruction training
- Developing a library of PSKM reference material
- Planning learning events for the successors
- Providing opportunities for successors to work alongside the experts
- Documenting Process Safety Knowledge and best practices
- Documenting effectiveness of critical technology such as incident management and action tracking tools

4.7 Establishing PSKM Culture

An organization with a strong PSKM culture addresses and tracks its process safety effectiveness. Learning from experience is shared from one office, region, or business line to another. It is important that multi-national organizations ensure messaging is consistent across the organization and corporate policies are implemented consistently throughout the organization. This practice reduces the risk of differing perceptions in an organization.

Organizational culture is the collective attitudes or behavioral characteristics which distinguish one organization or group of people from another. The culture of an organization is critical to the success of PSKM. An organization which fosters collaboration and learning is inherently designed to transfer knowledge and grow operational wisdom within its personnel. Consequently, if the organization fails to encourage networking, mentoring, exchanging of ideas and sharing of lessons learned as core values, it is ultimately ill-equipped to retain and manage the Process Safety Knowledge of its personnel.

To be successful, PSKM System implementation must consider the culture of an organization and describe how to communicate Process Safety Knowledge among its personnel so knowledge can be increased, updated, shared, and applied throughout the organization. Knowledge can be improved through process safety training requirements which can consist of, for example, process safety management fundamentals, chemical reaction hazards, high risk scenarios reviewed as part of hazard and operability studies, and hazards presented in area electrical classification drawings.

A success story in establishing an effective PSKM culture is described in Success Story 4.2.

Success Story 4.2 - Titan Industries

Titan, in 2006, identified two areas for continued success: knowledge management and sales [28]. Before implementing a knowledge management system, Titan's documents were stored and retrieved manually, and there was no collaboration workflow. It was difficult to find information and documents. Titan implemented a knowledge management system as a key component of its business strategy for improved product and service delivery to its customers. Titan utilized a technology platform as well as the development of custom software. All employees could share ideas, opinions, and knowledge independent of geographical location. Titan considered knowledge sharing to be a part of everyone's job. No rewards or incentives were created. The culture supported and expected it. Titan also formed Communities of Practice within key business areas. Titan's top success factor was their consideration of knowledge management as a defined business need. Titan also noted that there was strong support by the Chief Executive who was involved with the project development throughout the entire process. Titan developed a comprehensive range of metrics covering KM outcomes: deployment, use and effectiveness, though these were not strictly quantifiable.

4.7.1   Effect of Workplace Factors on PSKM

Technology, education, geographical location, and communication are examples of workplace factors that can impact PSKM.

Technology has an important role in the success of PSKM Technology makes it possible to improve the flow of information. If digital systems are outdated or software is insufficient or systems are not designed to accommodate the building of a digital PSKM system, then the implementation process can be significantly delayed.

Teams in different geographical locations require modern technology to communicate and share information. These team members may have varied backgrounds and may speak different languages. It is important that proper communication be established for effective interaction.

Communication also has an important role for sharing and transferring the PSK. Practices for sharing and transferring the PSK need to be evaluated to determine which methods are effective and which methods are not. Information shared and decisions made through emails may result in conflict if the information is not structured. Continual information in emails may lead to confusion for people who speak another language.

Examples of these workplace factors and their impact on PSKM implementation are presented in Table 4-10.

Table 4-10 Workplace Factor Effects on PSKM implementation

Workplace Factors	Effect on PSKM implementation
Technical	• Outdated digital systems, insufficient software or systems not designed to accommodate building a digital PSKM system can result in implementation delays. • Discrepancies can result when data exists in multiple systems which could lead to repeat work.
Educational	• Differences between people with academic qualifications could conflict with people who have progressed through an organization's ranks by on-the-job and/or skills training. Such differences could be between theoretical and practical viewpoints. • Style of communication may vary in different educational systems around the world. Some cultures place a high tolerance on other's viewpoints while others value cultural traditions.
Geographical	• Time zone differences can conflict with scheduling of virtual teams. • Long distances can prevent face-face interaction which is substituted with video conferencing or messaging.
Communication	• Information sharing and decision-making through emails may result in conflict if the information is not structured. • Poor communication or use of local jargon or slang may lead to misinterpretation and confusion. • Language barriers can create conflict between people if there are differences in native language, dialects, and accents, or use of different technical words or unfamiliar terms. • Network connectivity issues can interfere with the continuity of communication. • Verbosity of information in emails may lead to confusion for those who speak another language. • Using only one form of communication can be a barrier to sharing and receiving information.

4.7.2   Common Barriers to Sharing PSK

Implementation of an effective PSKM system leads to a knowledge sharing environment. However, there are barriers that may interfere with the flow of PSK. This is because people "bring their own values, beliefs, and habits into the workplace [33]." For this reason, it is important to understand the role that cultural factors may have related to the successful implementation of the PSKM.

For a successful implementation of PSKM, organizations need to implement solutions to identify and subsequently overcome these barriers.

Table 4-11 presents ten common barriers to sharing PSK in an organization [33] together with discussion and mitigation strategies. Each barrier has specific challenges.

Organizations need to address the barriers that pose the biggest concerns and develop strategies to drive the successful PSKM implementation forward. Removing barriers that inhibit PSK sharing will also establish a healthy PSKM culture that will aid in sustaining the PSKM implementation.

Table 4-11 Potential Barriers to PSKM and Strategies to Drive PSKM

	POTENTIAL PSK BARRIERS	EXPLANATION OF THE BARRIER	STRATEGY TO DRIVE PSKM AND ESTABLISH PSKM CULTURE.
1	Awareness	Unclear marketing of PSKM System leading to potential lack of understanding of PSKM accessibility and connection to PSKM experts	• Use communication platforms to promote PSKM and convey value of PSKM participation • Provide checklists and guidelines to help employees navigate PSKM
2	Cultural	Unwritten or spoken rules about how things are done	Align PSKM to company core values to make participation valuable to both employees and the organization
3	Distance	Geographic or structural separation resulting in infrequent contact	Increase availability or use of collaboration tools or conferencing platforms.
4	Experience (either very little or a great deal)	An employee with no substantial experience might feel they have little to offer and might not feel qualified to participate in knowledge sharing. An employee with a great deal of experience might prevent knowledge flow if they are considered an expert whose views are taken as definitive.	• Incorporate PSKM in on-boarding materials for new employees • Automatically enroll new employees into Communities of Practice • Provide opportunities for organization's subject matter experts to present and share their tacit knowledge and lessons learned to the work force.
5	Knowledge Hoarding	Hesitancy to share or contribute knowledge	• Recognize knowledge contributions. Structure advancement criteria to require those with subject matter expertise to leave a legacy by mentoring and training their successors as well as members of management • Establish a repository not held by one person

Table 4-11 Potential Barriers to PSKM and Strategies to Drive PSKM - Continued

	POTENTIAL PSK BARRIERS	EXPLANATION OF THE BARRIER	STRATEGY TO DRIVE PSKM AND ESTABLISH PSKM CULTURE.
6	Misaligned Measures	• Conflict of performance measures with knowledge-sharing behaviors between functions or business groups • Lack of measures or incentives to promote PSKM	• Include PSKM participation as part of a rewards/recognition system • Reference the Success Story 4.1 – Siemens AG presented in this chapter
7	Relationships	• Lack of connections between individuals • Lack of opportunities to communicate and network	Establish Communities of Practice to allow employees worldwide to interact, share information, and promote collaboration.
8	Sponsorship	• Failure to secure engaged sponsor to champion PSKM • Lack of alignment with core business objectives	Ensure commitment from senior leadership to convey importance of the PSKM and promote employee buy-in.
9	Time	• Viewed as a considerable time investment without corresponding benefits. • Inefficiencies such as asking subject matter experts to repeatedly answer the same questions.	• Ensure PSK content and expertise are available to employees at specific points in time to promote learning and knowledge reuse. • Ensure PSK is kept up-to-date especially following MOCs to existing processes or equipment.
10	Trust	Content is perceived to be unreliable because the sources of PSKM System are unknown or content is from former employees.	Develop networks of trusting relationships to bridge PSKM gaps.

Organizations must identify any barriers that prevent knowledge sharing and remove them. The main factors to overcome these identified barriers include [32]:

- Management or Executive Leadership
- Motivation, encouragement, and stimulation of individual employees
- Flat and open organizational structures
- Modern technology with a suitable sharing platform

4.7.3   Human Factors Effect on PSKM Development and Implementation

Human factors involve designing machines, operations, and work environments to consider human capabilities, limitations, and needs [34]. An organization's employees are key pieces of its processes. People are involved with gathering data and using it to generate information and knowledge which is ultimately used to make decisions. Because of the involvement of people, the likelihood of human error is always present. Understanding the design of PSKM and how it relates to human needs is discussed in this section. Table 4-12 [35] presents common human factors that address learning and application of knowledge.

4.7.4   Other Best Practices and Tools Related to a PSKM System

In addition to collecting data and information, knowledge management also involves how knowledge is stored, communicated, interpreted, and deployed for use within an organization. There are a variety of online world-wide knowledge management software and tools available for communications, messaging, and risk management. Software tools that support effective KM are recommended throughout the various stages of PSKM implementation. Types of software that can be used to manage PSKM are provided in Table 4 14.

Table 4-12 Effect of Human Factors on the PSKM System

HUMAN FACTOR	KEY POINTS	PSKM IMPACT
Information Processing Requires ability to process type and amount of information timely and effectively	• Consists of four stages: Sensing, perceiving, decision-making and action • Varies from person to person based on memory, attention, and decision-making capabilities	Make sure information falls within capabilities of employees as this will minimize failure to use, misinterpret, or forget the information
Attention Requires ability to focus while processing information Distractions are more likely to occur when fatigued, stressed or ill	• Ability to focus on a task at the right time • Requires full attention when completing unfamiliar tasks • Requires sounds and other signals to grab critical attention	In the workplace, people are occupied with many different inputs; information systems may require the addition of "attention grabbers" to refocus
Workload The chances of making a mistake may be higher for workers with too many tasks on their plates. Safety precautions may be overlooked leading to workplace accidents	• Ability to manage and process work • Requires an understanding of personal limitations during periods of high workload as well as situational awareness • Managing situational awareness is a continuous process that requires mental effort.	Prioritize information so that critical information is presented to the user, allowing them to process and respond quickly

Table 4-12 Effect of Human Factors on the PSKM System - Continued

HUMAN FACTOR	KEY POINTS	PSKM IMPACT
Decision-Making Potential to not see or misinterpret information. Inability to foresee the effect of changes	Decision-making is composed of: • Defining problem • Considering options • Selecting and implementing an option • Reviewing outcome Is influenced by availability of information and ability of the decision maker to interpret information, experience, social pressure, culture, and stress	Make Knowledge readily available in the PSKM System to facilitate decision making.
Stress Affects ability to process information and may lead to incidents	• Bodily response to a stimulus that disturbs or interferes with the "normal" physiological equilibrium of a person • Understanding of factors that lead to stress. Ability to cope with stressful situations can greatly improve an individual's performance	Develop PSKM System so Knowledge is well-organized and easy to find, thereby minimizing stress generated when looking for Information/Knowledge in critical situations.
Fatigue May lead to workplace accidents	• Probability of making errors increases when a person is fatigued • Signs of fatigue include reduction in attention, increased response times, small mistakes, and poor comprehension	Company policy regarding fatigue risks, limitation of shift work hours and shift rotation become knowledge

Table 4-13 Types of Software for PSK Management

SOFTWARE STAGE	TYPE OF SOFTWARE	PSKM TOOLS
Initial	Communication	• Document management for organization and storage of structured and unstructured information • Structured Information (i.e., sourced from sensors, network logs, online forms) and • Unstructured information (i.e., texts and social media posts) • Cloud or local network exchange of information
	Project Development	• Solution tools for • Advanced planning and scheduling • Quality and laboratory analysis • Process design • Mechanical design • PHA (HAZID/HAZOP/LOPA) • Bow Tie Analysis • Controls Review • Quantitative Risk Assessment • Risk Reports
Construction and Commissioning	Construction and Commissioning	• Vendor Design Package Information • Vendor Manuals • Operating Manuals and Procedures

Table 4-13 Types of Software for PSK Management - Continued

SOFTWARE STAGE	TYPE OF SOFTWARE	PSKM TOOLS
Operational	Risk Management	• Root Cause Analyses • Risk Identification and Assessment • Risk Registers
	Plant Operations	• Data Historian • Mechanical Integrity Records • Management of Change Records • Process Equipment and Technology Records • Predictive Component Failure Rates • Operational Excellence for Improvement and Reliability • Key Performance Indicators
	Incident Investigation / Tracking	• Incident Database • Closure of Incident Action Items
	Compliance Software	• Auditing
Communication	Communication Software	• Action Reports • Dashboards

These software types are examples that allow sustained and consistent application of PSK that employees can understand.

In-house registries and guides can be used to develop Hazard Registries, Incident Investigation Registries, Chemical Compatibility Charts, and Process Technology Guides.

Hazard registries can be generated during hazard reviews including PHAs, MOC reviews, and incident investigations [36]. Hazards can be listed by functional category. Table 4-14 presents example information that would be used to compile a Hazard Registry.

Table 4-14 Example Template Inputs for a Hazard Registry

Hazard Registry Template Input
1. Location of the Hazard
2. Brief Description of the Hazard
3. Risk Level of the Hazard
4. Risk Matrix Reference
5. Control Measures
6. Types of Controls

Incident investigation registries can be generated through compilation of data from incident investigations within a single plant or corporate-wide. These incident investigation registries can capture any common causes from incident investigations as well as learnings from audits and other management tools to identify gaps or human factors deficiencies. Table 4-15 presents a list of the information that can be used to build an Incident Investigation Registry template.

Table 4-15 Example Template for an Incident Investigation Registry

Metadata	• Plant/Process Location • Incident Data/Time • Incident Classification ((Incident, Near Miss, Unsafe Conditions, Unsafe Behaviors, Unsafe Actions Unsafe Inactions) • Process/Project Affected • Description of the Incident
Related Data for each "Type of Impact"	• Details of any injuries • Details of any equipment damage • Details of production loss • Details of environmental impacts • Details of (other impacts)
Affected Persons	• Investigation team leader • Investigation team members
Information Collection	• Relevant photos/videos of the incident • Location map/sketch • Operator logs • Potentially dozens of other documents
Investigation Results	• Identification of root causes • Follow-up recommendations
Closure	• Actions Taken

Chemical compatibility information can be compiled from safety data sheets and technical bulletins. Such information is useful for training, for preparing to conduct process hazard analysis studies, and for development of procedures including task specific personal protective equipment matrices. Most companies use a compilation of the safety data sheets, coupled with a reactivity grid and material compatibility grid as the chemical hazard registry. The CCPS tool "Chemical Reactivity Worksheet" or "CRW" can be used to develop chemical reactivity and material compatibility [37].

Another example of an in-house registry is the process technology guide listing known consequences of and best practice safeguards for multiple scenarios. These registries can be referenced during hazard review studies to make sure that teams address identified critical issues. Similarly, the registry should be updated if the hazard review identifies previously unidentified scenarios.

The same information can be applied for updating safe operating limits or mechanical design limits, and consequence of deviation tables.

4.8   Chapter Summary

In summary, this chapter provides detailed discussions of the PSKM System development and implementation. It includes key steps and examples to develop a PSKM System and potential obstacles to overcome during system implementation and subsequent sustainment of the system. A process for capturing and organizing the knowledge is presented together with potential barriers to the PSKM along with strategies to drive the PSKM and establish culture. Tools for data management are also discussed.

4.9   Introduction to the Next Chapter

In the next chapter, maintaining and improving the PSKM system is introduced. One of the most challenging areas in PSKM is ensuring the knowledge and wisdom developed within people is maintained and not lost when a person leaves the organization or changes their position within the company. Successfully addressing maintenance and improvement challenges for the PSKM System requires both a well-designed program and dedicated people supporting this program. Key topics to be discussed in the next chapter will include:

1. Managing Organizational Change
2. Auditing
3. Performance Monitoring and Indicators for the PSKM System

5   Maintaining and Improving PSKM

"The field of knowledge management (KM) was created to help control processes by providing just-in-time information." Lyle C. Emmott

Once the PSKM process has been developed and implemented, the next step is to maintain and improve the process to ensure it remains robust. The continuous improvement of PSKM is achieved mostly through new understandings, new opportunities, or revisions in the information. If new processes are added, initial PSK will need to be developed. Connection to MOC, project management, and R&D must be in place and effective. Data analysis can be used to determine if the results of a change make a difference.

When knowledge is generated, it must be determined where it will be stored, used, and maintained. Each requires a procedure that explains these details so that it is clearly understood. Information for Process Safety Knowledge (PSK) can be collected from company data systems and global sources. The following list is an example of records and source references:

- Original Equipment Manufacturer Manuals
- RAGAGEP
- Design Calculations based on Worst Case Scenarios
- Government Requirements
- Equipment Standards
- Managements of Change
- Safety Data Sheets
- Process Hazard Analyses

The organization can then use the information collected as a guide for the development of measurements and key performance indicators (KPIs) to benchmark the performance and identify improvement opportunities. Benchmarking is important to drive continuous improvement and to ensure that the program remains effective, accurate, and as complete as possible. This information will drive process wisdom when it is used for:

- Proposing managements of changes
- Preparing for PHA studies
- Conducting process audits
- Investigating incidents for root cause identification

In Section 5.1, roles, and responsibilities for managing PSK are presented. In Section 5.12, the tools needed to maintain PSK are explained. These tools are intended to provide maximum benefit to employees. The tools are used to ensure PSK is accurate and reliable so that when used by employees they can trust the knowledge is relevant and usable for the process. The fact that employees can then rely on the PSK, means they will actively engage in maintaining it. Employees will more readily engage with the PSKM system if a company has a healthy process safety culture combined with a policy that directs employees to use the PSKM System. A policy by itself may not bring about increased use of the PSKM System if the information in the system is not accurate and reliable. An improvement cycle then results, with employees using the tools because they are accurate and reliable and employees updating information and knowledge in the tools to keep them accurate and reliable. In Section 5.3, tools used to measure the effectiveness of the PSKM System are presented for comparison of actual results to target goals. This comparison will help identify improvement opportunities to reach optimum PSKM performance.

5.1   People, Roles and Responsibilities for Managing the PSK

Resources are required to maintain the PSK. Clearly defined roles and responsibilities enable management to identify the types of qualified people needed to successfully maintain the PSK. Examples of qualifications include certifications, industry and company experience, skills, and knowledge.

Table 5-1 presents an example of RACI chart for maintaining PSKM. According to this example, the Chief Knowledge Officer (CKO) is accountable for maintaining the PSK in the organization. The responsibility of maintaining the PSK falls within the responsibilities of PSKM Stewards. This structure allows for the proper distribution of the responsibilities within the whole organization and requires regular consultation from the other PSKM roles to support the full maintenance of the PSKM system.

Table 5-1 Example PSKM RACI Chart for Maintaining Knowledge

| Role / Job Title | Mentoring | Leads PSKM Project | Knowledge | | | |
			Capturing	Obtaining	Maintaining	Providing
Chief Knowledge Officer (CKO)	A	A	A	A	A	A
PSKM Project Manager	C	R	C	C	C	C
PSKM Champions	R	I	C	C	C	C
PSKM Navigators	C	I	C	C	C	R
PSKM Stewards	C	I	R	C	R	C
PSKM Editors	C	I	C	R	C	C

(R) Responsible; (A) Accountable; (C) Consulted; (I) Informed

Knowledge resources that are available to organizational teams to maintain the PSK include examples in Figure 5-1. Especially, a well-maintained library of lessons learned is critical for a sustainable PSKM system. Some examples for building and maintaining such a library is presented in Section 6.2.

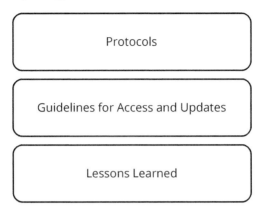

Figure 5-1 Knowledge Sources

5.1.1   **Logic Model for Maintaining PSKM**

As introduced in Chapter 4.1.1, a logic model can be used as a framework to maintain the PSKM system.

Table 5-2 presents example inputs and activities to maintain the knowledge. Inputs to this table are outputs from Table 4-8. While some of the outputs are also commonly generated as part of doing business, others may not be, and therefore must be explicitly generated. Some data management tools incorporate elements of the outputs. Management of Change procedures should be extended to knowledge maintenance practices. MOC (both Personnel Change and Operational Change) is integral to maintaining the PSK and the PSKM System.

Table 5-2 Example Logic Map Components to Maintain Knowledge

		MAINTAIN THE KNOWLEDGE		
INPUT SOURCE	**CONSTRAINTS**	**ACTIVITIES**	**OUTPUTS**	**EFFECTS**
Information Type: CHEMICALS / PROCESS CHEMISTRY / PROPERTIES AND HAZARDS				
Hazardous effects of inadvertent mixing Information communicated	Supplier chemical composition differences. SDS updates. Training materials not up to date	Audit/revise program as needed to address changed conditions (e.g., new chemicals, new hazards). Update training plans to include: • Hazardous effects of Inadvertent mixing • PPE Matrix	Process reviewed for obtaining and reviewing SDS to determine if new procedures need to be developed to handle chemicals and update training.	Procedures and training materials updated as needed. Employees informed of any chemical hazards they may encounter while on the job.
Information Type: EQUIPMENT DESIGN AND OPERATING LIMITS				
Computerized maintenance management system (CMMS) populated	Time Commitments, Resources to manage CMMS updates for new equipment and daily maintenance activities.	Identify roles and responsibilities, Audit system.	Responsible functions for CMMS updates identified. Audits implemented.	Maintenance design and operating limits information centralized and updated when needed.

Table 5-2 Example Logic Map Components to Maintain Knowledge – Continued

	MAINTAIN THE KNOWLEDGE			
INPUT SOURCE	**CONSTRAINTS**	**ACTIVITIES**	**OUTPUTS**	**EFFECTS**
Information Type: INCIDENTS AND NEAR MISSES				
Risks categorized, tracking system for resolution implemented.	Poor decision-making may result if records are not updated and actions closed.	Track and confirm resolution status in management meeting agendas.	Incident management system improved.	PHA studies updated with updated risk and resolution information. PHA teams able to make good decisions with updated information.

5.2   Tools to Maintain the PSK

PSK is an important element of successful employee training. Accurate and complete process data, information, and knowledge are necessary. The information can be applied daily within the process or manufacturing area and used to understand process hazards and analyze risks. To gain knowledge, it is necessary to be able to understand the process safety information.

Organizations maintain a combination of both static (slowly, infrequently, or never changing) and dynamic information. Static information is typically fixed but may change over the lifetime of a facility. On the other hand, dynamic information continually changes after it has been recorded. In PSK, dynamic information uses static information to influence process incident trending and evaluation, PHAs, MOCs, operating procedure development, and training programs. The dynamic information dictates how PSK is utilized and manipulated to better serve the organization. PSK can be seen as both a "static response to an inquiry" as well as "dynamic knowledge supporting an ability to respond to unpredictable situations." [38]

Table 5-3 Examples of Static and Dynamic Information

Static Information	Dynamic Information
• Chemical physical properties • Safety Data Sheets • Product and Equipment Specifications* • P&IDs* • Operating Procedures* • Equipment design • Process safety information • Reaction Chemistry • Incident Investigation Reports • Incident History	• Company Dashboards • Learning from incidents • Software that monitors and records process parameters such as temperature, pressure, and level • Key Performance Indicators • Number of knowledge searches with rank (provides where the users found the information) • Skills Mentors, number of mentor-mentee pairs • Knowledge Mapping • Knowledge Management Training • PSKM Training progress • Documenting Process Safety Knowledge and best practices • Development of PSKM reference material • Process Audit findings • Knowledge capture interviews • Standards in process safety (RAGAGEP)

* Product and Equipment Specifications, P&IDs, and Operating Procedures can change over the course of a facility's life, as changes are made to the process or as operating experience is gained.

5.2.1   Technical Training and Employee Development Program

Employee development programs are critical for organizational success. It is important that employee development programs involve people across an organization. Development programs will reduce the dependency on tacit knowledge, reduce risks and facilitate decision-making. Properly documented and maintained procedural and work processes will help ensure that work is performed consistently and will enhance knowledge transfer [39].

Training enhances an individual's knowledge, skill, and/or competency in a process safety critical position that can directly influence the prevention of and recovery from major incidents. Process Safety Management elements are interrelated and PSM training lays the foundation for targeting incident and hazard reduction. Such foundational education training can be applied to job related hazard scenarios for the purpose of gathering data and information to generate best practices and for process safety decision-making. Scenario-based training along with knowledge of an organization's safety systems are building blocks of information used to develop best practices for implementation and sharing.

Most people want to grow and to feel more competent and more responsible, at any level. Employees who are given opportunities to update their skills are inclined to remain with the organization. There is job satisfaction when employees have opportunities to use their skills and abilities at work [40].

Personnel training education can be a key factor in delivering PSK to site personnel to apply within the process area. Continuous training education on lessons learned and refresher training on operating and safe work practice procedures can help to sustain PSK of site personnel and, combined with experience, it can transform knowledge into wisdom. Process Safety certification programs offered through AIChE / CCPS, as well as other chartered engineering programs, provide the latest developments for people to build PSK competencies.

A formal mentoring program is a key process for an organization to retain and transfer its knowledge. Mentor / mentee programs establish two-way communication for key learning, knowledge retention and fostering risk awareness. Mentoring is an effective way of passing down the knowledge from one individual expert who has the knowledge and wisdom to another who seeks the information to acquire the knowledge. The goal is for an organization to retain its expertise and develop additional expertise. This can be accomplished through a knowledge development plan that incorporates a targeted timeline and key assignments under the supervision of the mentor, face-to-face meetings, training education, progress reports and assessments.

5.2.2   Management of Change (MOC)

MOC is a PSM management system to identify, review, and approve all modifications to equipment, procedures, raw materials, and processing conditions, other than replacement in kind, prior to implementation to help ensure that changes to processes are properly analyzed (for example, for potential adverse impacts), documented, and communicated to employees affected [1].

MOOC is a framework for managing the effect of new business processes, changes in organizational structure or cultural changes within an organization.

An organization's MOC program or procedure should describe how change is managed within an organization. The program should consist of a database for properly documenting changes, required action items to implement the change safely and hazard reviews / approvals that were conducted.

MOC is a critical tool to ensure that all the documentation affected by a change is captured and maintained, accessible for review at any time or in PHAs, and reinforces or helps drive the overall PSKM program. Documenting changes to a process, ensures that the PSK is updated every time a change occurs.

A mature MOC program protects the process safety design by conducting robust risk assessment and hazards identification and by updating all process safety information involving the chemical hazards, technology, and equipment.

5.2.3   MOC to Maintain and Update Existing PSI

MOC is present at the beginning and at the end of the PSKM system. Existing equipment/process modification MOCs serve to manage existing documentation or add information to existing PSI as changes are made.

A robust Management of Change (MOC) or best practices for MOC program can be adopted to document changes, update Process Safety Information and associated PSK, and ensure changes are effectively communicated to the stakeholders. Changes need to be effectively communicated to all people impacted by the change including all shifts that will operate the changed process, maintenance, technical and management personnel. It takes effort for an individual to "update" their mental model, the thinking tools used to make decisions and solve problems, and for the mind's eye to see an internal picture of how a new process works. The more sources a person must draw upon, the clearer the thinking becomes [41]. This aspect is important to sustain PSK.

The MOC process should be rigorously followed by an audited process, demonstrating to employees that new and updated knowledge is an asset. When employees view knowledge management as value-added, they are more inclined to use it.

Companies spend hundreds of hours documenting hazards and controls in their Process Hazard Analysis (PHA) studies. This information is critical for PSK to be formally documented. By making the commitment to properly maintain PSK, employees will view knowledge management as a core value.

5.2.4   Incremental Changes

Process variables may vary slightly over time. These changes may be subjective to human factors such as following steps in an operating procedure. An operator may not recognize them as a change because of their gradual change, but these small changes should be captured in HAZOPs dealing with sequence or order of operations changes, through annual operating procedure review and certification, and annual review of a plant's standard operating conditions and limits. Any deviations identified should be corrected and documented to bring the process back to its original design.

An organization's PHA process must be dynamic or inclusive enough to capture any subtle changes that have occurred over time. This can be done by identifying changes since the previous HAZOP and ensuring that the associated hazards are known and adequately managed. This is particularly important to ensure that design safeguards continue to be adequate. A thorough HAZOP should include documentation and discussion of all changes that have occurred to the process since the last study and a "revalidation" of setpoints and procedures of operation.

5.2.5   Hazard Registry – Risk Matrix

Hazard registries and risk matrices are tools used to build and sustain PSK. These hazard studies and matrices aid in the generation of like or similar deviation-cause pairs for identification of high-risk or undesirable consequences that may be common to manufacturing sites with similar processes and manufacturing chemistries.

The identification process leads to the compilation of large consequences, such as fires, explosions, or air dispersion of hazardous materials, together with the respective cited safeguards and recommended mitigation plans. These mitigation plans could be used for further development and implementation of best practices across different sites with similar processes. Hazard studies and matrices aid in the identification of high-risk or undesirable consequences which are broadly classified as human, environmental or economic impacts [42].

The bulk of the PSK is documented in PHA studies. It is of utmost importance that the information regarding deviation / cause / consequences scenarios be clearly described in as much detail as possible to truly understand what could happen and what is going to prevent it. For example, instead of documenting "fire" as a consequence, it could be described as "a pool fire fed from a 500-gallon vessel of flammable solvent." Another example is instead of documenting a consequence of "pipe failure", it could be described as "pipe corrosion at the flange connection at the bottom of the vessel". For the PSK to be useable, people need to understand what is occurring or failing to cause the deviation and how bad it could be.

A hazard registry-risk matrix could be used by manufacturing plants when they revisit their PHA, typically on five-year cycles. This process would generate the knowledge and subsequent wisdom for managing high-risk hazards. A

Hazards Register database could be developed by documenting all the pertinent information related to the risks assessed. The database could also contain the resolution of each hazard and not only the latest resolution but the history starting from the original study [36].Mitigation plans could be used for further development and implementation of best practices across different sites with similar processes.

In a global company, corporate process safety groups should develop safety management guides (SMGs) to be shared across the organization. One example is a guidance document for conducting and participating in Hazard and Operability Studies (HAZOPs) to drive consistency in terms of allocation of severity, frequency, and identification of safeguards. This will lead to the creation of a knowledge base of high-risk scenarios, scenario risk estimates, and adequacy of safeguards across an organization. These guidance documents will also serve as training tools for PHA leaders and teams.

5.2.6   Process Safety Communities of Practice

The organization should have a plan or method to develop Subject Matter Experts (SMEs). The SMEs provide guidance and advice on internal guidelines, RAGAGEP, risk tolerance, and overall setting the standard for PSK within the company. They should be involved in internal (and possibly external) mentorship programs to help develop or pass on PSK to the next SMEs. They should also be actively involved in external process safety organizations, conferences, and other forums to assist with development of industry wide PSK.

Organizations should also form process safety communities of practice composed of SMEs, such as Chemical, Mechanical, Controls, Environmental Engineers, Process Operators, Industrial Hygiene Specialists and Process Safety Engineers. Process safety communities of practice participants should come from various manufacturing sites that are the most familiar with the organization's high hazard scenarios and materials. These SMEs could develop and maintain consistent hazard prevention and safeguarding guidelines. The SMEs should periodically connect at internal or external conferences, roundtable meetings or other platforms, to exchange information, review existing knowledge and further improve the guidance for distribution within the organization.

A large organization may well have the resources to provide process safety communities of practice upon demand, but not all companies are large enough to do so. Smaller companies, or companies working within a smaller supporting infrastructure (such as companies operating within developing countries) may not have the ability to staff a community of practice from within. However, professional associations, non-governmental trade organizations, governmental trade departments and 3rd party engineering and process safety companies may be used to generate sufficient interest and support to fulfill the functions of a community of practice.

5.2.7   RAGAGEP for Equipment Specific Information

"Recognized and Generally Accepted Good Engineering Practices (RAGAGEP) are documents that provide guidance on engineering, operating, or maintenance activities based on an established code, standard, published technical report, or recommended practice (RP) or a document of a similar name." [43]

A company should make available RAGAGEP materials to personnel within the organization in addition to any internally developed standards, best practices, guidelines, or regulations. An employee should be able to easily access internal standards (internal RAGAGEP) and be able to determine (through references in the internal standards) the external standards that serve as the basis for the internal standards.

RAGAGEP combines wisdom developed by industry experts to provide information and knowledge in the form of codes and standards. These standards are important information for design and installation of process vessels, equipment, and instrumentation and to establish inspection frequencies and appropriate testing methodologies.

A manufacturing company that documents the RAGAGEP it will rely on, such as API 510, Pressure Vessel Inspection Code, will ensure that the relevant appropriate information and knowledgeable personnel oversee inspections, repairs on vessels and pressure-relieving devices. Maintenance of codes and standards by knowledgeable personnel will preserve the information and ensure equipment integrity. Adopted RAGAGEP should be summarized and maintained for knowledge reference. Examples of RAGAGEP are shown in Tables 9-1 through 9-3 in CCPS Guidelines for Mechanical Integrity Systems [43].

RAGAGEP can also be generated internally as standards, best practices, and engineering design guidance documents.

Access to codes and standards provides engineers and operations personnel opportunities to interact with specialized individuals to exchange information about specific designs and provide "easier pathways to validated information." [44]

Guidance and learnings from technical trade associations on specialized topics are sources of Process Safety Knowledge maintained by those standing committee members who collaborate and update industry standards. These trade organizations are credible sources that keep industry up to date on technical and safety related information and emergency response procedures as well as any regulatory changes. Examples of such specialized sources and topics are:

- The Chlorine Institute or Eurochlor (in Europe) for Chlor-Alkali safety
- CCPS Golden Rules [45]
- OSHA (U.S.), INERIS (France), HSE (U.K.) Fact Sheets covering hazardous chemicals
- The International Institute of Ammonia Refrigeration (IIAR) professional society providing ammonia refrigeration best practices
- The Ammonia Safety & Training Institute and the European CEFIC umbrella organization pertaining to first responder training
- The Compressed Gas Association (CGA), European Industrial Gas Association (EIGA) and the Asian Industrial Gas Association (AIGA) developing guidelines for safe design and operation of industrial gas facilities, plus safe transportation and handling of industrial gases and cryogenic liquids

5.2.8   Maintenance and Inspection Historical Data

Maintenance personnel develop and maintain knowledge from multiple sources:

- Technical documents such as maintenance procedures and troubleshooting guidelines
- Classroom experience gathering sessions
- Computer-based training
- Intentional and planned on-the-job training working with a subject matter expert
-

PSK can be maintained through use of maintenance management software tools, technical information, maintenance management systems and inspection data as summarized in Table 5-4 [46]. Personnel assigned to maintain and repair critical equipment need to be able to interpret data to understand and correct negative trends and to perform recommended procedures to avoid losses.

Table 5-4 Tools and Methods for Maintaining Maintenance-related PSK

Maintenance Management Tools	PSKM Maintenance Resources
Maintenance Management Software Tools	Equipment troubleshooting guides Operation Manuals Equipment Schematics and Diagrams Asset Information Preventive Maintenance Scheduling Tools
Inspection Reports	Inspection Results and History Performance Data Historical Equipment Photos
Supporting Procedures and Manuals	Procedures and related maintenance tasks Equipment Installation and Operation Operation and Maintenance Manuals Managements of Change
Equipment Technical Files	Equipment Type Design and Construction Information Equipment Repair History Inspection, testing, and preventive maintenance activities Vendor-Supplied Information
Metrics and Quality Assurance (QA) Data	Quality Assurance Tasks Detailed instructions for QA tasks Identification of equipment needing maintenance
Safety Device Functionality	Proof Testing Required Frequencies Testing History and Inspection Records Testing Criteria
Incident Metrics	Incident equipment-related causal factors Events and near misses involving equipment

The two major components of maintenance data are Inspection Records and Maintenance Records. Inspection Records contain information on equipment, maintenance, failures, spares, and consumable materials. Maintenance Records contain information about staff and budgets. Maintenance managers mainly focus on failure type and rate, maintenance actions, breakdown repair costs, and performance measurements.

Data and information stored in computerized maintenance management systems can be fully utilized to extract and obtain knowledge. Trends in equipment failure rates can be extracted to gain knowledge for preventive maintenance scheduling and for making continuous improvements. Capturing patterns of knowledge is known as Knowledge Discovery in Databases. This process can be used to generate new knowledge. By applying knowledge discovery in databases, the next potential failure can be predicted [47].

Historical data can be used to predict potential equipment failure for decision-making to avoid loss events [47].

5.2.9   Shared Learning Database

To sustain PSK, it is important that companies maintain an incident database. This is an important tool that can be used to extract information about past incidents and their initiating causes and root causes. The tool can also be used to identify trends and major contributing root causes and causal factors. The database should contain documentation of incident details, initial risk assessments, investigation team members, investigation results, recommended actions to address root contributing and causal factors, final risk assessments after recommendations are in place, and communications of learnings within the organization. The incident database is only as good as the PSK it contains.

Some industry trade associations such as the American Fuel & Petrochemical Manufacturers (AFPM) and the American Chemistry Council (ACC) also maintain incident sharing databases for member access.

Maintaining an incident database provides benefits to PHA Teams and other manufacturing plants with similar processes or process chemistries. PHA teams reference the incident database when studying cause-consequence scenarios. Manufacturing plants with similar processes or process chemistries reference the incident database for root cause analysis and best practice development if corrective measures are populated within the database for each incident.

Two global sources that can be referenced to obtain information regarding incident investigations and lessons learned are the Center for Chemical Process Safety (CCPS) and the U.S. Chemical Safety and Hazard Investigation Board (CSB). CCPS has developed a monthly one-page newsletter called the Process Safety Beacon [24] to share important process safety incidents and lessons learned for manufacturing personnel. The U.S. Chemical Safety and Hazard Investigation Board investigates root causes of major chemical incidents. CSB's searchable incident database contains current investigation progress and completed investigations with root causes and issued recommendations to protect people and the environment [25].

5.2.10   Corrective Action Tracking System

Corrective action information collected from internal/external audits, safety assessments, incidents, maintenance, or inspection deficiencies and other hazard reviews improve the Process Safety Knowledge. Corrective actions are intended to eliminate root causes and eliminate undesirable consequences. Preventive actions are intended to prevent a potential problem from recurring. The information provides a clear picture from which an organization can measure any improvements or identify repeated process safety issues and areas that need correction.

An organization that develops corrective actions from an investigation can use that knowledge to address potential weaknesses in other areas. For example, if an investigation identified the omission of a processing step in an operating procedure as a causal factor, then the corrective preventive action could be to review the remaining operating procedures for the same omission.

By focusing on the most significant problems and implementing preventive and corrective actions, the organization will continually improve. For every investigative accident or incident, corrective action plans should be established. Root causes and initiating causes should be categorized to identify patterns and common cause failures.

Finally, corrective action plans should be audited to evaluate the effectiveness of the corrective actions. This information should be documented in the database to provide overall knowledge of the history of incidents, root causes, corrective actions, and effectiveness.

5.2.11   Document Management

Document management is a critical process for organization and maintenance of PSK documents. Both electronic and information that is not electronic need to be managed.

5.2.11.1 Electronic Records

Databases would serve to electronically house the organization's critical documents such as:

- Operating Procedures
- Piping and Instrumentation Diagrams (P&IDs)
- Process Parameters
- Inspection Reports
- Maintenance Records
- Incident Reports
- Chemical Compatibility Information

The database must be organized to be easily accessible and retrievable when needed, maintained through the MOC process, updated when changes are made, easily searchable (keywords, titles, authors), and provide full text information.

Such a database could reside on an organization's intranet homepage as a "Document Library" showing the tool in a single view and categorized by document type with drop down menus to provide document title, owner, purpose, last update, and revision history.

5.2.11.2 Paper Records

Paper records need to be protected, too. These records can include inspection records, original equipment manuals, checklists, redlined prints, and similar items that need to be organized and managed appropriately. Organizations need to ensure that their document management procedures for paper records are just as rigorous as management procedures for electronic copies. Essential components for building a successful document management plan for paper records include:

1. Performing an inventory audit and creating a filing system to be able to retrieve the documents quickly.
2. Archiving any documents that need to be kept for legal purposes.
3. Scanning paper documents and storing them digitally in a database.

Training should include records management to ensure that employees understand the procedures for managing records securely.

5.3   Assessing and Improving the PSKM Program

To be successful in PSKM, organizations need to continually monitor their programs. By developing and implementing key performance indicators (KPIs), organizations can continually measure and assess the effectiveness of their PSKM. Measurements provide reliable data. The data must be compared to anticipated results. Baseline measurements that are initially captured should be trended, and then compared to previous results to determine if performance is improving. The comparison will provide insight into what aspects are working or not working within the PSKM program.

The insight will help an organization understand the results of the measurements. The organization needs to understand the results of the measurements and convert measurements into action plans for improving the PSKM. This process is known as performance management. Performance measurement will track progress, but the results of the measurement will need to be managed. Performance measurement and performance management will help an organization become and stay successful in PSKM.

In addition, a successful PSKM program can use five components: People, Processes, Technology, Structure, and Culture. These five components are described in more detail in Table 5-5.

5.3.1   PSKM KPIs

KPIs are used to gauge performance, evaluate success, and assess variances when compared with targeted goals. The relationship between KPIs and Targets is presented in Figure 5-2. The PSKM KPI is compared to the targeted goal or desired state. KPIs are used to measure the health of the PSKM system and its components.

PSK KPIs can be used to monitor performance, usage, and drive improvements. KPIs can also be used to identify trends from which performance action plans can be identified to improve the program. Positive trends signal progress or success. Negative trends signal an alert that the measured PSK variable is trending away from the targeted goal. KPIs that can be used to monitor the health of the PSKM are provided in Table 5-6.

Table 5-5 Core Components of a Successful PSKM Program

CORE COMPONENT	PSKM CONSIDERATIONS FOR LONG-TERM SUCCESS
PEOPLE	Develop skills and competencies so that teams can be productive in the areas of active listening, communication, and critical thinking.
PROCESSES	Develop best practice documents to identify, manage and spread the knowledge.
TECHNOLOGY	Verify tools can be used to enable users to access and update the PSKM.
ORGANIZATIONAL STRUCTURE	Make certain that organizational structures maintain PSKM across disciplines and expertise.
CULTURE	Confirm that the organization fosters continued growth and maintains PSK sharing.

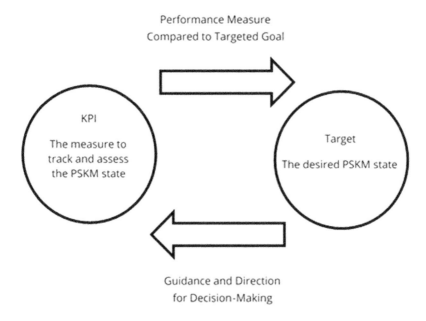

Figure 5-2 Performance Measurement and Performance Management

Organizations should select indicators to measure the performance of specific business and Process Safety Knowledge goals. KPIs provide targets for organizations to gauge progress over time and help people make decisions on courses of action needed to drive improvement. KPIs that indicate below target performance identify areas of weakness for which improvement is needed. KPIs should be reviewed regularly to ensure continued improvement and once improved, should be adjusted to the next area(s) of concern.

In Table 5-7, KPIs for measuring PSKM performance are provided. There are five types of knowledge management indicators that can be used to measure an organization's PSKM Performance. Organizations should select applicable indicators and develop scorecards, charts, or dashboards to monitor progress.

Table 5-6 Suggested PSKM Health Measurement KPIs

PSK KPI	PSKM Health Measurement
PHA Completion Status	Number of PHA studies completed on time and any significant PHA completion gaps or delays
PHAs with complete Process Safety Information available	Example: (number of PHAs with complete Process Safety Information (PSI) available / number of PHAs performed) x 100%
PHA Recommendations	Number of recommendations are completed on time compared to total open recommendations
Incident Investigation	Number of repeat incidents, incidents involving similar equipment, and corrective actions timely completion; process knowledge gaps which contributed to the incident
Maintenance	Inspection, Testing, Preventive Maintenance timely completion rate
Audit Results	Identification of high-risk corrective measures, timely completion rate of corrective actions
MOC Status	Frequency of changes occurring to a process, action items timely completion, timely closure, tracking of document updates and communications to impacted individuals
Employee performance ratings and progress through development plans	Targets being met, does the KPI suggest more resources or opportunities are needed

Table 5-7 PSKM Performance Indicators

Type of KPI	KPI Insight
PSK Contributions	• Provides analysis of PSKM contributors and contribution frequency. KPIs help determine whether the organization is growing the PSK database • Positive Trends indicate that the workforce is contributing to the knowledge base • Negative Trends indicate decreasing PSK contributions for which performance management actions are needed to reverse the trend
PSK Interactions	• Provides analysis of which types of PSK are target searches • Positive Trends indicate the importance of the PSK content • Negative Trends may indicate that the PSK lacks importance or significance, or users may not be aware of its value which would initiate performance management actions
Response Time	• Provides question response time, user engagement, and question resolution • Positive trends may indicate quick resolution and that the PSK users are able to quickly access information or receive a response • Negative trends may indicate that information was difficult to find or that there was no or slow resolution for the information which would initiate performance management actions
Account Utilization	• Provides information on number and frequency of use of the PSK database • Positive trends may indicate high activity rates or high percentage of use by the organization • Negative trends may indicate that PSK information is difficult to find, may not be relevant to current needs, may not be up-to-date or the percentage of use is dropping which would initiate the need for performance management actions.
Search Activity	• Provides information on frequency and topics of PSKM searches • Repetitious searches may indicate that information is difficult to find for which performance management actions are needed to make the information easier to locate or the information is important or of higher priority

5.3.2   PSKM Maturity Assessment

A Maturity Model is a tool that an organization can use to assess its PSKM effectiveness and identify capabilities and improvements needed to move to higher levels of maturity. Levels range from low or initial levels where knowledge management might be a concept to a fully integrated ideal system. As an example, a four-stage PSKM Maturity Model may be separated into levels of increasing complexity on a scale of 1-4. Level 1 is the lowest rating and level 4 is the highest rating or best in class. Level 1 PSKM System is considered immature and is defined as Reactive. Level 2 PSKM System meets the minimum standards or requirements set by the company or regulation and is defined as Dependent. Level 3 PSKM System can demonstrate improvement over time and is defined as Independent. Level 4 PSKM System focuses on continual improvement and is defined as Optimized.

Table 5-8 shows an example of a four-level Maturity Model.

Table 5-8 Example Four Level Organizational Maturity Model

Level 1	Reactive	The organization's knowledge is conceptual, siloed, finding information is difficult.
Level 2	Dependent	Knowledge repository meets minimum standards.
Level 3	Independent	Improvements over time can be demonstrated.
Level 4	Optimized	All employees are ambassadors of the system, who contribute to, utilize, and share the system; success can be demonstrated continuously.

The assessment is first conducted to determine the current level of an organization's PSKM maturity. The assessment would be protocol driven to include interviews and document review based on a set of sampling criteria. Once the maturity level is assessed, then a target level above that is chosen as a goal to attain with set criteria to meet. Priorities can be set to identify the capabilities and actions needed to reach a fully mature system.

The maturity markers can be defined in terms of the elements of Capture, Organize, Maintain and Provide. As presented in Chapter 4, the framework outlined can be used to gauge PSKM maturity levels as shown in Table 5-9.

Table 5-9 Example Logic Map Components to Gauge PSKM Maturity Level

INPUT EXAMPLE	Reactive	Dependent	Independent	Optimized	Outputs
"CAPTURE" PSKM MATURITY LEVELS					
RESPONSIBILITIES BY ROLE	Organization has roles and responsibilities but not necessarily for PSKM.	Organization's internal standards are in preliminary stages.	Organization has a written procedure / format for developing internal standards but has not published internal standards for PSKM.	Organization has published internal standards for PSKM and shared with Staff. A RACI Matrix has been developed.	RACI Chart
"ORGANIZE" PSKM MATURITY LEVELS					
RACI CHART	Responsibilities for PSKM are not communicated.	Team overwhelmed with unnecessary communication.	Organization has a developed RACI matrix but has not communicated it to all levels.	Organization has approved a RACI matrix. All personnel understand and utilize it. Teams communicate clearly.	Maintain clear and open communication

Table 5-9 Example Logic Map Components to Gauge PSKM Maturity Level - Continued

INPUT EXAMPLE	Reactive	Dependent	Independent	Optimized	Outputs
		"MAINTAIN" PSKM MATURITY LEVELS			
MAINTAIN CLEAR AND OPEN COMMUNICATION	PSKM goals are not established.	PSKM goals have been developed but not communicated between Leadership and Front-Line Workers.	KPIs have been developed but have not been rolled out to all levels within the organization.	All levels within the organization are aware of PSKM goals. KPIs are current and presented at regular intervals.	Clearly defined PSK goals
		"PROVIDE" PSKM MATURITY LEVELS			
CLEARLY DEFINED TEAM PSK GOALS	There is no organizational structure established for PSKM.	Knowledge is being captured but is scattered or there is duplication of effort since it is not well organized.	Organizational goals and structure have been defined.	Organization has shared its goals with all levels of the organization with frequent KPI updates.	Productivity and faster feedback

5.3.3 Gap Assessments

Using the PSKM Maturity Framework, a gap assessment can be conducted to identify elements that are missing in the process. In Chapter 4, gap assessment was identified as a requirement under Organizational Structure to determine what knowledge is needed by the organization, what knowledge is available, who has the knowledge or who can provide it. Gap assessment can also be extended to include the skillsets and knowledge needed by the organization. The skills and knowledge assessment will identify which employees have the most knowledge of the company's business as well as any employees that have important skills gaps. This gap analysis will determine what training is needed or how to bridge the gap or evaluate new hires.

To complete a skills gap analysis, critical competencies and skills must be identified and assessed. This can be accomplished by identifying the PSKM goals, creating an inventory of skills or information needed, inventorying the skills and information already developed, and comparing skills and information needed against skills and information possessed.

A gap assessment will help identify the current state of the PSKM and the gaps between the current state and the desired state. Performance management actions will close any gaps.

Table 5-10 provides examples of gap assessment questions to capture, organize, maintain, and provide knowledge in the same context as the Maturity Model discussed above.

Table 5-10 Example Gap Assessment Questions

Capture	Organize	Maintain	Provide
Does the organization know what knowledge to capture? Are roles and responsibilities for PSKM defined?	Has a RACI chart for PSKM been developed?	Has the organization developed a succession plan for critical PSKM roles?	Is the RACI chart communicated? Do people understand their roles in PSKM?
Is the PSKM embedded in the minds of people or documented?	Is the PSK organized in a location that can be accessed by all people who need it? Are SMEs identified in company network systems?	Is the relevant PSK periodically reviewed to ensure that the data and information is current, and correct? Is the knowledge of SMEs kept up to date? Is the company active in identifying and developing future SMEs?	Is the new PSK communicated to the workforce including SMEs? Are revised or discontinued changes in PSK communicated to the workforce?

5.3.4   PSKM Audits

A PSKM audit differs from a regulatory audit. A regulatory audit focuses on the availability of documents and records maintained by the organization against established criteria. A PSKM audit expands on a regulatory audit such that it covers not only availability of documents but their content, accuracy, system/process to create/update, and how the information is shared and utilized. The PSKM audit provides valuable feedback to the organization covering:

- Information storage and maintenance
- Information accessibility
- Whether the information is current and relevant
- Knowledge gaps
- How the information is shared across the organization
- MOC process – are changes being translated into the PSK and communicated?

A PSKM audit benefits an organization by identifying gaps in the system and improvement opportunities. Seven benefits that can be realized by a PSKM audit are:

1. A better understanding of how PSK flows and propagates across an organization. PSK can flow in parallel, serial, or coupled patterns. The flow of PSK can identify areas for improvement.
2. Gaps in the PSK
3. Siloed PSK that could be of benefit to others within the organization.
4. PSKM storage requirements
5. Duplicate or differing PSK
6. Shared view of the PSK that is available across the organization
7. Effective MOC and MOOC (both process changes and organizational changes)

Prior to initiating a PSKM audit, a knowledge inventory should be developed to identify what information and knowledge exists in the organization, where it is stored, and whether it is explicit or tacit knowledge. PSK can exist on company intranets, shared drives, or can be uncovered during subject matter expert interviews, questionnaires, and surveys.

An evaluation of PSK inventory can help identify:

- Information duplication issues
- Information that resides in multiple locations
- Buried information
- Siloed PSK that is not available to others within the organization
- Obsolete or incorrect PSK

Once PSKM systems are established, the systems will need to be audited to ensure continued effectiveness. In addition, sites should establish KPIs to monitor the health of the overall system and identify potential issues early. Historically auditing of Process Safety Knowledge has focused on the documentation. However, to audit PSKM, an organization must also audit the level of understanding within the key personnel who interact with and support a process. Having an auditing process in place ensures the Process Safety Knowledge remains accurate and complete, but it must incorporate learnings from events as well as changes made over time.

The effectiveness of the audit depends on the auditor's skills, experience, and knowledge. Auditing of Process Safety Knowledge Management can be performed in several ways by a variety of different resources. Audit teams may be internal or external to the organization or a combination of both. Types of audit teams are presented in Table 5-11. Internal audits are often referred to as first-party audits, while external audits can be either second-party or third-party.

Table 5-11 Types of Audit Teams

Type of Audit	Performed By	Definition
First Party Audit	Audit Team Internal to the Site	Site performs the review of PSKM using internal resources against established criteria.
Second Party Audit	Audit Team Internal to the Company but external to the site	Company personnel from outside the site perform the review of PSKM using internal resources against established criteria.
Third Party Audit	Audit Team External to the Company	External company, contractor or consultant performs the review of PSKM using external resources against established criteria. A third-party audit is free of any conflict of interest and provides independence.

It should be noted that there is complementary fit between PSM compliance audits and PSKM audits. The ability to provide the right knowledge to the right person at the right time during a PSM compliance audit is a KPI for the PSKM System. This observation is especially true as information becomes more and more relegated to silos by data management systems.

5.3.5   Preparing for the PSKM Audit

Table 5-12 provides a list of twelve components of a knowledge audit. From this list, audit criteria can be established to perform a PSKM audit.

While preparing for a PSKM audit, there are many audit resources that should be consulted. These audit resources are listed in Figure 5-3. These audit resources along with 12 common elements of a knowledge audit help establish a PSKM Audit Protocol. Once the initial PSKM Audit Protocol is established, it can be continuously improved after each PSKM audit.

Table 5-12 Knowledge Audit Criteria

NO.	ELEMENT	AUDIT CRITERIA
1	Document Quality	Are documents readable and in standard format?
2	Knowledge Processes	• What percentage of projects or incidents concluded with a "Lessons Learned" development that was shared with company personnel? • How many incidents cited "lack of knowledge" in the process incident description?
3	Knowledge Culture	• Is knowledge hidden in silos? • What is the percentage of documents private to each team that contribute to hidden knowledge in silos?
4	Usability	Do user interfaces for PSKM impact productivity or produce poor quality search results that contribute to knowledge waste?
5	Knowledge Loss	• Do exit interviews indicate that knowledge is leaving the organization without being captured? • Have mentoring programs been developed and implemented? • How many mentoring programs are currently active?
6	Knowledge Waste	Are there any instances where employees could not locate and/or access knowledge in a timely and efficient manner?
7	Knowledge Overhead	Have time and cost been assessed for production of low value documents?

Table 5-12 Knowledge Audit Criteria - Continued

NO.	ELEMENT	AUDIT CRITERIA
8	Information Security	Are there instances where knowledge has been sent by email attachments instead of by secure link?
9	Document Control	Have individuals ignored a requirement to approve or acknowledge digital documents and other knowledge artifacts?
10	Structure and Organization	• Have statistics of knowledge repositories and tools been evaluated to decide the average depth of a document in a user interface? • Are there any average depths greater than four levels that could contribute to knowledge waste by making documents difficult to find?
11	Knowledge Velocity	Are valuable documents going unused or are documents being created at a significant expense that are of low value?
12	Document Analytics	Has an evaluation been made of broad statistics related to use of documents to identify waste and value?

Auditing Resources

There are several good resources which can be used including:
CCPS Guidelines for Risk Based Process Safety
- Chapter 8 Table 8.1 Typical Types of Process Knowledge
- Section 8.3.2 Catalog Process Knowledge in a Manner that Facilitates Retrieval
- Section 8.3.3 Protect and Update the Process Knowledge
- Chapter 14 Training and Performance Assurance

CCPS Guidelines for Auditing Process Safety Management Systems
- Chapter 9 Process Knowledge Management
- Chapter 15 Training and Performance Assurance
- Appendix A: PSM Audit Protocol
- CCPS Process Safety Metrics Guide for Selecting Leading and Lagging Indicators version 3.2 April 2018 section 7.5.2.1 Process Hazards Analysis (PHAs)
- CCPS Guidelines for Risk Based Process Safety – Chapter 8 Section 8.5 Metrics
- AIChE G-27 – CCPS: Guidelines for Process Safety Documentation

---

Figure 5-3 Auditing Resources

5.3.6   PSKM Software

Knowledge management software is often available as an addition to standalone resource planning systems. Each organization must choose software solutions that meet their needs. Whether the organization is looking for new software or to improve existing software, there are essential requirements that each organization must evaluate to ensure that the software they are selecting meets their PSKM needs. Nine items often used for software evaluations are presented in Table 5-13.

Table 5-13 Knowledge Management Software Evaluation Criteria

ITEM	SOFTWARE REQUIREMENT	EVALUATION CHECKLIST
1	Deep Index Search	Evaluate if the search engine is user-friendly and powerful enough to index all words across all document types including spoken words in videos.
		The benefit is that it allows users to quickly find information needed even if they do not know in which folder it is stored.
2	Quick Access to Essential Content	Evaluate if navigation tools or links to access PSK can be prominently displayed on company websites or internal homepages.
3	End User Segmentation	Evaluate if the software is capable of customizable notification settings to control the flow of relevant PSK or allow users to opt into relevant conversations and mute conversations that are not relevant.
4	Cross Publication of Content	Evaluate if the PSK can be shared across communities of practice and is accessible and up to date from no matter which community it is generated from.
5	Customized PSKM Interface	Evaluate if the software can provide the ability to customize the PSKM homepage or communications to provide consistency and a seamless experience using the tools.
6	Multiple Ways to Identify and Categorize Users	Evaluate if users can be categorized by region or teams or can create reports specific to the user's needs.

Table 5-13 Knowledge Management Software Evaluation Criteria – Continued

NO.	Software Requirement	Evaluation Checklist
7	Additional Reporting Available Through Application Programming Interface	Evaluate if the software allows applications that can communicate with each other. This capability allows administrators to easily find and review, optimize, clean, or remove outdated materials. Provides ability to find out users that have made modifications to the information, remove outdated information, and optimize information.
8	Integration with Email or Chat	Evaluate if users can comment or answer email or notification questions or messages.
9	Customized Reporting	Evaluate if the software can provide specific reporting requirements such as dashboard analytical data.

5.4   Information Management

Information is an asset and should be cared for as other business critical items.

5.4.1   Information Management Technologies

Three sets of information technologies can be used to manage data:

1.   Communication Technologies
2.   Collaboration Technologies, and
3.   Storage and Retrieval Technologies.

5.4.1.1   Communication Technologies

Communications protocol must contain clear instructions for each communication type (i.e., For Informational Purposes, Review, or Action Required). While everyone receives email, to ensure good communication actually occurs, people need to be aware of the action they need to take (if any) based on the receipt of that communication.

Communication technologies allow users to access needed information and communicate with each other. The communication process is both one-way and two-way. Examples are verbal, visual, email communication, company intranets, the internet, forums, and other web-based tools that provide communication capability. There is no preferred method of communication for PSKM. However, the strategy that the PSKM team uses must be easy to understand and shared in real time. A study conducted in 2008 by researchers with the University of Johannesburg [48] found that the respondents from participating organizations, when asked to single out one method deemed most effective for organizational communication, chose email as the most preferred method over face-to-face, cell phone, telephone, and other methods. Many companies discourage the use of email for sharing information and knowledge and prefer to see employees provide links to the information management system where the specific info/knowledge is stored.

5.4.1.2   Collaboration Technologies

Collaboration technologies provide the means to perform group work and electronic brainstorming, even across geographic distances. An example of a collaboration technology is a meeting software platform that allows multiple users to work on the same files. Electronic brainstorming uses computer technology to enter and automatically circulate ideas in real time to all group members.

5.4.1.3   Storage and Retrieval Technologies

Storage and Retrieval technologies use database management tools to manage knowledge. The challenge for organizations is to identify the technologies needed to meet their PSKM needs.

Document management systems need to be easy to use, accessible to all employees in all locations, and work on numerous software platforms, including computers, cell phones, and other smart devices. Systems can either involve document retrieval, database systems or reference retrieval.

Document retrieval systems store entire documents. These documents are normally retrieved by title or key words associated with the document. Text may be stored as data which permits full text searching. Retrieval may be made based on any word that is in the document [49].

Database systems store information as records that can be searched for and retrieved based on content. The data are stored by the computer for ready access [49]. Reference retrieval systems store references to documents rather than the documents. This type of system is efficient for storing large amounts of printed data [49]. The main characteristics of a storage and retrieval information system are presented in Figure 5-4 [50].

Figure 5-4 Characteristics of a Storage and Retrieval Information System

Data reliability requires preservation of data, whether it is for incident investigation or preserving records of operations, operator logs, and maintenance history. These records need to be stored in easy-to-use database systems or data historians. Safety critical data requires systems for preservation. Consider the characteristics for data preservation in operations environment presented in Figure 5-5 [51]. The taxonomy of a document system for a process plant must follow the company's organization to be useful [52].

Protect data transmission from field to control room and data management (data historians) from potential hazards (fires, heavy rains, flooding, dropped objects and electromagnetic or radio frequency interference)	Arrange safety critical data in a format that is easy to use for investigations and troubleshooting
Timestamp the Data	Ensure Data Security
Use modular systems to facilitate system expansion	Perform system upgrades to avoid obsolescence of the data management system
Implement administrative safeguards to reduce probability of accidental destruction of information	Implement data backup protocols and decide how long data will be maintained in the historian and frequency of data capture (i.e., every second or every 5 seconds)

Figure 5-5 Characteristics for Data Preservation in Operations Environment

5.5   Chapter Summary

In summary, one of the most challenging areas in PSKM is to ensure that the knowledge and wisdom developed within people is not lost when people change their roles within the organization or leave the organization. Successfully meeting this challenge requires both systems and people. This chapter provided detailed discussions of the PSKM System maintenance and improvement. Chapter 5 further provided tools to maintain Process Safety Knowledge Management and KPIs to improve the PSKM Program through relevant PSKM KPIs and PSKM Audits.

5.6   Introduction to the Next Chapter

In the next chapter, case studies in PSKM are presented and reviewed.

6   Case Studies and Lessons Learned

"It might seem to an outsider that industrial accidents occur because we do not know how to prevent them. In fact, they occur because we do not use the knowledge that is available [53]." Trevor Kletz

6.1   Introduction to PSKM Focus Charts

This chapter includes lessons learned from chemical process safety incidents involving ineffective or missing Process Safety Knowledge Management (PSKM) systems. These lessons are presented as Case Studies. Each study is organized in three parts: 1) an incident summary, 2) an incident PSKM Focus Chart summarizing PSKM System failures, and 3) benefits of a PSKM System had it been in place at the time of the incident. While the overview summarizes the incident, the accompanying Focus Chart depicts the causes and their relationship to the PSKM System.

Note that committee's selected "excerpts"---direct quotes using paragraphs from the publicly-available incident investigation reports---are used to help explain and summarize the case study. At the time that the case study report was issued, the investigators did not have "PSKM systems" in their terminology. Although many readers may be familiar with the case study and its report, the reader is invited to re-read each case study in this guideline, including its excerpts, with PSKM systems in mind.

A PSKM Focus Chart is used to highlight the relationships between the PSKM System elements and various factors related to the specific incident discussed in the Case Study. The first Case Study is used to provide the step-by-step development of a Focus Chart.

The following terms are applied in the PSKM Focus Chart:

PSKM Focus Chart - A chart divided into three columns that depict causes and other factors related to the incident, and four rows that show elements of the PSKM System (i.e., Capture, Organize, Maintain, and Provide).

Cause (Incident) - An event, situation, or condition which results, or could result (Potential Cause), directly or indirectly in an accident or incident [1].

Contributing Cause - Factors that facilitate the occurrence of an incident such as physical conditions and management practices. (Also known as contributory factors) [1].

Proximate Cause - The cause factor which directly produces the effect without the intervention of any other cause. The cause nearest to the effect in time and space [1].

Root Cause - A fundamental, underlying, system-related reason why an incident occurred that identifies a correctable failure(s) in management systems. There is typically more than one root cause for every process safety incident [1].

In the context of the PSKM Focus Chart, Contributing Cause, Proximate Cause, and Root Cause are related to the PSKM System.

This chapter also includes key factors from PSKM success stories in other industries. The chapter concludes with risk analyses of common causes of knowledge management system failures and countermeasures for success.

6.2   Case Studies from Significant Incidents

The first case study discussed in Section 6.2.1 is used to describe the step-by-step development of a PSKM Focus Chart. All remaining case studies present the incident's completed PSKM Focus Chart.

6.2.1   BP Texas City, Texas

The following incident description includes information excerpted from several sources. The incident will be reviewed using the principles of the PSKM system introduced in previous chapters. This framework will enable identification of PSKM System factors identified as proximate, contributing or root causes that were either present or absent, and how these factors contributed to the outcome of the incident.

1) Incident Summary:

On March 23, 2005, at 1:20 p.m., the BP Texas City Refinery suffered one of the worst industrial incidents in United States (US) history. Explosions and fires fatally injured 15 people and injured another 180, alarmed the community, and resulted in financial losses exceeding $1.5 billion. The incident occurred during the startup of an isomerization (ISOM) unit when a raffinate splitter tower was overfilled, pressure relief devices opened, releasing flammable liquids to a blowdown drum and ultimately to the atmosphere, instead of routing them to a flare for safe destruction (Figure 6-1). The release of flammable materials led to an explosion and fire. The fatalities occurred in or near temporary office trailers located close to the blowdown drum. A shelter-in-place order was issued that required 43,000 people to remain indoors. Houses were damaged as far away as three-quarters of a mile (1.2 km) from the refinery.

The BP Texas City facility was the third-largest oil refinery in the US. Prior to 1999, Amoco owned the refinery. BP merged with Amoco in 1999 and subsequently took over the operation of the plant. Based on the CSB Report [54] published after the accident: "BP had process safety systems in place in its Group management system and at Texas City, yet in the wake of the merger with Amoco, the resulting organizational changes to safety management led to a de-emphasis of major accident prevention."

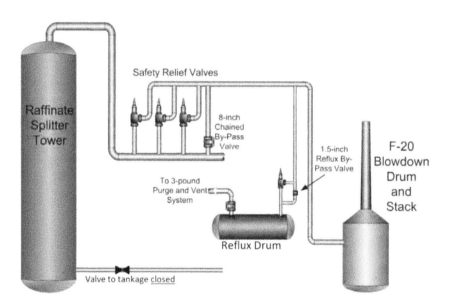

Figure 6-1 BP Texas City Raffinate Splitter Tower and the Blowdown Drum

2) PSKM Focus Chart:

Based on the published Investigation Report [20], the underlying causes were identified as:

"Over the years, the working environment had eroded to one characterized by resistance to change, and lacking trust, motivation, and a sense of purpose. Coupled with unclear expectations around supervisory and management behaviors, this [deterioration] meant that rules were not consistently followed [including startup procedures], rigor was lacking and individuals felt disempowered from suggesting or initiating improvements."

"Process safety, operations performance and systematic risk reduction priorities had not been set and consistently reinforced by management."

"Many changes in a complex organization led to the lack of clear accountabilities and poor communication, which together resulted in confusion in the workforce over roles and responsibilities."

"A poor level of hazard awareness and understanding of process safety on the site resulted in people accepting levels of risk that were considerably higher than comparable installations. One consequence was that temporary office trailers were placed within 150 feet (46 meters) of a blowdown stack which vented heavier than air hydrocarbons to the atmosphere without questioning the established industry practice [and performing a thorough hazard analysis]."

"Given the poor vertical communication and performance management process, there was neither adequate early warning system of problems, nor any independent means of understanding the deteriorating standards in the plant."

There was clearly a breakdown in the PSKM system in this incident. PSKM causes that played a role in the failure of the PSKM System failure included the following:

PSKM Proximate Cause:

The Raffinate Splitter Tower level was not observed or known during startup resulting in liquid overfill of the tower.

PSKM Contributing Causes:

1. Operators did not follow startup procedures by sending flammable liquid hydrocarbons into the Raffinate Splitter Tower for over three (3) hours without any liquid being removed since the level control valve on the tower was closed by an operator. This failure led to flooding of the tower and high pressure. The essential failure in the PSKM system was the operators did not understand the negative consequences of not

following start-up procedures including correctly setting and verifying the current positions of critical valves during the Startup.

2. Control instrumentation provided false level indication and critical alarms failed to work as designed. The operator did not recognize the control system failure since it did not indicate a mass imbalance between liquid added and removed. These failures were contributing causes of the incident but not necessarily a contributing cause for the failure in the PSKM system. The specific failure for the PSKM System was that operators either did not know or did not communicate that level control instruments could not be relied upon since they were prone to failure.

PSKM Root Causes:

1. ISOM operators were likely fatigued from working 12-hour shifts for 29 or more consecutive days. The National Institute for Occupational Safety and Health (NIOSH) describes fatigue and its consequences as slower reaction time, reduced attention or concentration, limited short-term memory, and impaired judgement [55]. A fully functional PSKM system relies on operators that are not fatigued. Moreover, there was an ineffective knowledge transfer between shifts. The start-up was a hazardous phase of the operation which needed additional support and attention. The failure related to the PSKM System was that the management did not provide experienced and technically trained personnel who would communicate the hazards during unit startup.

2. Operator training program was inadequate. The fact that control instrumentation might provide false level indication and critical alarms might fail to work as designed was not conveyed in the training. Although shift handover was discussed during new operator training, no structured guidance was given, preventing adequate assessment of trainees' ability to effectively communicate between shifts.

3. Recurring operational problems resulted in normalization of deviation and reinforced the perception that procedures did not have to be followed during startup. These recurring operational problems were not captured, organized, or may have been excluded from operator training program.

The proximate cause "The Raffinate Splitter Tower level was not observed or known during startup resulting in liquid overfill of the tower" is the subject of this PSKM Focus Chart. Separate PSKM Focus Charts can be built for other proximate causes. For instance, another proximate cause for the PSKM System failure could be "Placement of temporary office trailers within 150 feet (46 meters) of a blowdown stack venting hydrocarbons heavier than air to the atmosphere".

The first row of the PSKM Focus Chart is constructed by relating the proximate cause to the contributing and root cause to illustrate the gaps that existed in the PSKM System for providing knowledge at the time of the incident (Figure 6-2).

	PROXIMATE CAUSE	CONTRIBUTING CAUSES	ROOT CAUSE
PROVIDE KNOWLEDGE	The Raffinate Splitter Tower level was not observed or known during startup resulting in liquid overfill of the tower.	Operators did not understand the negative consequences of not following startup procedures including correctly setting and verifying the current positions of critical valves during the startup.	The management did not provide experienced and technically trained personnel that would communicate the hazards during unit startup.

Figure 6-2 BP Texas City PSKM Focus Chart - Provide Knowledge Gap

The outdated and ineffective procedures led operators to believe procedures could be altered or did not have to be followed during the startup process. BP Texas City management did not ensure the startup procedure was regularly updated even though modifications to the unit's equipment were made over time. Management also allowed operators to make procedural changes without performing proper Management of Change. Management did not ensure operational problems were corrected over time which led to operators deviating from the established procedures.

The operator training program did not include a formal program for crews to discuss potentially hazardous conditions such as startup or shutdown to enhance operator knowledge or training for abnormal situation management. Training did not include discussion of the potential failure of control instrumentation and critical level alarms. Because these topics were not included in operator training, the operators were not aware of how to recognize malfunctioning instruments. Operators were also not aware of the consequences that could occur if the instrumentation was going beyond its range. Although shift handover was discussed during new operator training, no structured guidance was given, preventing adequate assessment of trainees' ability to effectively communicate between shifts. Moreover, malfunctioning instrumentation did not alert operators to the actual conditions of the unit. During the March 23, 2005, startup, the level transmitter indicated the liquid level in the splitter tower was gradually declining, although it was in fact rising. Operations personnel involved in the raffinate section startup were unaware the transmitter's reading was inaccurate.

Incident investigation revealed the day shift supervisor arrived late for his shift and did not conduct a shift turnover with any night shift personnel thus failing to capture information regarding startup activities from the previous evening. Although the Day and Night Board Operators did speak to each other at shift change, the Night Board Operator was not able to provide much detail about the previous night's raffinate section startup activities since he was not the person who filled the tower. That person had left the refinery approximately an hour before his shift ended and did not add much detail in the logbook. An effective shift hand over, however, involves both the verbal exchange of information as well as capturing the salient points in a written form.

The second row of the PSKM Focus Chart describes different causes that led to failures in maintaining the knowledge as shown in Figure 6-3. BP Texas City management did not implement findings from previous incident reports that recommended more training for operators and supervisors, the use of control board simulators, and the use of fully functional operating conditions and limits. Moreover, reports of malfunctioning instrumentation were not acted on and instrument checks were not completed prior to ISOM startup. BP Group's "Getting Health, Safety, and the Environment Right" (GHSER) policy was established in 1997 to provide a business wide HSE management system. The GHSER expectations encompassed both personal safety and some process safety elements, however, GHSER reporting requirements focused on personal safety. The GHSER policy did not require BP's Refining and Marketing segment, including oil refining, to report explicit process safety performance indicators such as unclosed action items from PHAs or incidents.

	PROXIMATE CAUSE	CONTRIBUTING CAUSES	ROOT CAUSE
PROVIDE KNOWLEDGE	The Raffinate Splitter Tower level was not observed or known during startup resulting in liquid overfill of the tower.	Operators did not understand the negative consequences of not following startup procedures including correctly setting and verifying the current positions of critical valves during the startup.	The management did not provide experienced and technically trained personnel that would communicate the hazards during unit startup.
MAINTAIN KNOWLEDGE		Operators either did not know or share that level control instruments could not be relied upon since they were prone to failure.	Operator training program was inadequate.

Figure 6-3 BP Texas City PSKM Focus Chart - Maintain Knowledge Gap

The CSB Report [54] concluded "...operators biased the tower level on the high side, likely to avoid the possibility of losing the level and damaging the furnace. This bias of running the tower on the high side occurred in nearly all of the last 19 startups. The operators' actions show they were aware of the tower's low-level risks, but not the high-level risks." Based on these factors, the third row of the PSKM Focus Chart can be built to capture failures in organizing knowledge as illustrated in Figure 6-4.

	PROXIMATE CAUSE	CONTRIBUTING CAUSES	ROOT CAUSE
PROVIDE KNOWLEDGE	The Raffinate Splitter Tower level was not observed or known during startup resulting in liquid overfill of the tower.	Operators did not understand the negative consequences of not following startup procedures including correctly setting and verifying the current positions of critical valves during the startup.	The management did not provide experienced and technically trained personnel that would communicate the hazards during unit startup.
MAINTAIN KNOWLEDGE		Operators either did not know or share that level control instruments could not be relied upon since they were prone to failure.	Operator training program was inadequate.
ORGANIZE KNOWLEDGE		Operators were biased towards avoiding low level in the tower instead of potential issues with high levels.	BP Texas City did not create an effective reporting and learning culture.

Figure 6-4 BP Texas City PSKM Focus Chart - Organize Knowledge Gap

Before the Texas City 2005 incident, a similar incident occurred at the Texaco Milford Haven Refinery in the United Kingdom in 1994 (described next). That incident resulted in an explosion following a plant upset. The direct cause of the explosion was a combination of failures in management, equipment, and control systems. The lessons learned from the UK incident were shared with BP in 1997 [54]:

"A high distillation tower level led to a fire and explosion incident reported by the U.K. Health and Safety Executive (1997). In July 1994, at the Texaco

Milford Haven refinery, a plant upset resulted in liquid hydrocarbons being pumped into a distillation tower for several hours while liquid flow out of the bottom was shut off. This problem was undiagnosed by operations personnel, as the instruments indicating the outlet valve position, level, and flow out of the tower bottom provided inaccurate information. As a result, the tower over-pressured and caused a release of flammable hydrocarbons when outlet piping from the flare knock-out drum failed. The lessons from this incident were addressed to the U.K. refinery and petrochemical industry, including BP (HSE, 1997). The Health and Safety Executive found that the operators were not provided adequate information on process conditions and that Texaco was not adequately monitoring instrument maintenance and equipment inspection."

"In its report on the Milford Haven explosion, the U.K. Health and Safety Executive recommended an automatic safety device such as the shutdown of tower feed to prevent overfilling (HSE, 1997). The Instrument Society of America has suggested high level overrides on column feeds to reduce, and eventually halt, the flow of material into a column if the liquid level is high (Buckley et al., 1985). For columns where the consequences of overfilling could be severe, redundant instrumentation and a high reliability safety instrumented system (SIS) are recommended. Such systems, which operate independently of the normal operating controls, improve the reliability of safeguards, and help reduce the likelihood of catastrophic failures such as the March 23, 2005, incident."

Based on the CSB Report on the BP Texas City incident, BP's system for capturing learnings from other incidents was ineffective. The full PSKM Focus Chart depicting the PSKM System gaps that occurred during capturing, organizing, maintaining, and providing knowledge is provided in Figure 6-5.

	PROXIMATE CAUSE	CONTRIBUTING CAUSES	ROOT CAUSE
PROVIDE KNOWLEDGE	The Raffinate Splitter Tower level was not observed or known during startup resulting in liquid overfill of the tower.	Operators did not understand the negative consequences of not following startup procedures including correctly setting and verifying the current positions of critical valves during the startup.	The management did not provide experienced and technically trained personnel that would communicate the hazards during unit startup.
MAINTAIN KNOWLEDGE		Operators either did not know or share that level control instruments could not be relied upon since they were prone to failure.	Operator training program was inadequate.
ORGANIZE KNOWLEDGE		Operators were biased towards avoiding low level in the tower instead of potential issues with high levels.	BP Texas City did not create an effective reporting and learning culture.
CAPTURE KNOWLEDGE		Similar recurring operational problems were not recognized as deviations from normal operations.	Lessons learned from Texaco Milford Haven Refinery in the United Kingdom in 1994 were not captured.

Figure 6-5 BP Texas City Case Study PSKM Focus Chart

3) PSKM Benefits

An effective PSKM system could have helped BP prevent the Texas City Incident. The safety lessons learned from the Texaco Milford Haven fire and explosion that occurred in 1994 during startup were shared with the U.K. refinery and the petrochemical industry, including BP, who acquired the Texas City Refinery after merging with Amoco in 1999. If an effective PSKM system had been in place at BP at the time of the incident, then the safety lessons learned from the Milford Haven incident could have been shared with the people who operated the Texas City Refinery, potentially avoiding the incident at Texas City.

Both Texaco's and BP's incidents were similar in several ways, such as occurring during startup. The Texaco Milford Haven incident was a result of an over-pressured distillation column and a release of flammable hydrocarbons. Liquid hydrocarbons were pumped into a distillation tower for several hours while liquid flow from the bottom was shut off. The problem was undiagnosed by operations personnel. Instrumentation indicating valve position, levels, and exit flow provided inaccurate information. The Health and Safety Executive [20] found the operators were not provided adequate information on process conditions and Texaco was not adequately monitoring instrument maintenance and equipment inspection.

During the Texas City Refinery startup, operations personnel pumped flammable liquid hydrocarbons into the tower for over three hours without any liquid being removed, which was contrary to instructions provided in the startup procedure. Critical alarms and control instrumentation provided false indications that failed to alert the operators of the high level in the tower. Consequently, unknown to the operations crew, the 170 feet (52 meters) tall tower was overfilled, and liquid overflowed into the overhead pipe at the top of the tower. BP Texas City Refinery's operator training program was found to be inadequate. The central training department staff had been reduced from 28 to eight, and simulators were unavailable for operators to practice handling abnormal situations, including infrequent and high hazard operations such as startups and unit upsets.

There were parallels between the Texas City and Milford Haven Refinery incidents, yet the Process Safety Knowledge developed following the Milford Haven incident that was shared with BP was not captured nor provided to the operating staff at the Texas City Refinery. A more robust PSKM system within BP could have helped prevent the Texas City Refinery incident.

This case study presents an in-depth review of the PSKM system consisting of the Capture, Organize, Maintain and Provide model and demonstrates the importance of providing necessary knowledge so that everyone who needs it, has it and at the right level of detail.

Figure 6-6 Destroyed trailers near ISOM unit at BP Texas City

Source: [54]

6.2.2 Concept Sciences (CSI) Hydroxylamine Explosion, Allentown, Pennsylvania

"On the day of the incident, CSI was producing its first batch of 50 wt-percent HA solution at the new facility." [56]

1) Incident Summary:

At 8:14 pm on February 19, 1999, a process vessel containing several hundred pounds of hydroxylamine (HA) exploded at the Concept Sciences, Inc. (CSI), production facility near Allentown, Pennsylvania. Employees were distilling an aqueous solution of HA and potassium sulfate, the first commercial batch to be processed at CSI's new facility. After the distillation process was shut down, the HA in the process tank and associated piping explosively decomposed, most likely due to high concentration and temperature. Four CSI employees and a manager of an adjacent business were fatally injured. Two CSI employees survived the blast with moderate-to-serious injuries. Four people in nearby buildings were injured.

The following process description is excerpted from the U. S. Chemical Safety and Hazard Investigation Board (CSB) Case Study [56]:

"CSI's distillation process included a 2,500-gallon (roughly 9,500 liters) charge tank which was 25 feet (7.6 meters) long and 4 feet (1.2 meters) in diameter; a vacuum distillation system, which consisted of a glass column (heating column) and remote water heater, a glass condenser (condenser column) and remote chiller, and a vacuum pump; and two product receivers consisting of a forerun tank and a final product tank, both 1,500-gallon (roughly 5,600 liters) tanks, 15 feet (4.5 meters) long and 4 feet (1.2 meters) in diameter. The distillation is performed in two phases. The first phase of the process begins as a pump circulates the 30% (by weight) HA from the charge tank to the heating column, a vertical tube-in-shell glass heat exchanger. The HA enters the top of the column and is heated by 120 degrees Fahrenheit (about 50°C) distilled water as it cascades through the tubes back to the charge tank. Vapor from the column is condensed using a chilled water condenser (condenser column). The distillate, initially consisting primarily of water with some HA, is directed into the forerun tank (Figure 6-7) [56]."

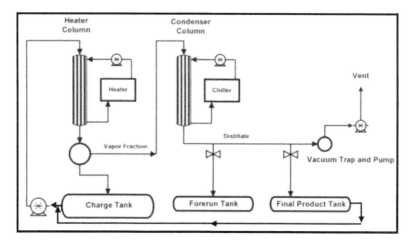

Figure 6-7 CSI Vacuum Distillation Process

"When the concentration of HA reaches 10% in the forerun tank, the distillate is diverted to the final product tank, where it is collected until the concentration of the liquid phase in the charge tank is 80 to 90% HA. At this point, the first phase of distillation is complete. The charge tank and column are cleaned using a 30% HA solution, and the charge tank is taken out of service. In the second phase of distillation, the 45 weight-percent HA solution collected in the final product tank is further concentrated by redistillation. It is fed back to the top of the heating column and flows through the tubes, where it is heated by 140°F (60°C) water. The distillate is directed back to the final product tank. Water is removed from the HA solution until the material in the final product tank reaches 50 weight-percent HA, at which point the distillation is complete."

The following excerpt is from the incident Technical Report as published by the U.S. Federal Emergency Management Agency [57]:

"The explosion registered 0.7 on the Richter Scale at Lehigh University's seismograph center, which is located five miles from the CSI Plant in Bethlehem, Pennsylvania. The University's readings indicate that the explosion occurred at 20:14:43 hours on Friday, February 19, 1999, and that the blast caused the ground to move both up and down and side to side. The explosion was seen as far as seven miles away and was felt as far away as Lehighton and Tobyhanna to the north and Trexlertown and Longswamp Township to the west."

2) PSKM Focus Chart:

The U. S. Chemical Safety and Hazard Investigation Board (CSB) specifically addressed the issues of inadequate process knowledge and documentation. The CSB [56] concluded the following:

- At CSI, the development, understanding, and application of process safety information during process design was inadequate for managing the explosive decomposition hazard of HA.
- During pilot-plant operation, management became aware of the fire and explosion hazards of HA concentrations more than 70% (by weight). This knowledge was not adequately translated into the process design, operating procedures, mitigative measures, or precautionary instructions for process operators.
- CSI's HA production process, as designed, concentrated HA in a liquid solution to a level of more than 85%. This concentration is significantly higher than the referred 70% concentration at which an explosive hazard exists.
- Only sketches and basic process flow diagrams were developed; there were no standard engineering drawings.
- Operating procedures provided only rudimentary information.
- Engineering drawings and detailed operating procedures should have been a key component of operations and maintenance training.

PSKM causes that led to the PSKM System failure in this incident consist of the following:

PSKM Proximate Cause:

HA was distilled to an 85% concentration, above the safe concentration limit of 70% and operations staff were not aware the process operated at unsafe concentration levels.

PSKM Contributing Causes:

1. The fire and explosion hazards of HA concentrations more than 70%, as documented in the Safety Data Sheet was not adequately translated into the process design, operating procedures, preventive measures, or precautionary instructions for process operators.
2. Operations staff were not aware of the potential hazards.
3. The potential for runaway crystallization was never identified.
4. No hazard scenarios were captured.

PSKM Root Cause:

4. Basic process safety and chemical engineering practices were not adequately implemented.
5. The process was not designed to control the reactive hazards.

The PSKM Focus Chart in Figure 6-8 depicts the gaps to capture, organize, maintain, and provide Process Safety Knowledge.

3) PSKM Benefits:

The incident at CSI may have been less likely to occur with an effective PSKM system in place. Following the incident, the CSB found CSI had deficiencies in:

6. Process Knowledge and Documentation
7. Information Management
8. Process Safety Reviews for Capital Projects
9. Reactive Hazard Reviews
10. Process Hazard Analysis

These five items are all components of an effective PSKM system. Had PSKM System been implemented, CSI could have captured adequate process safety information and provided the specific fire and explosion hazard information to the people who were developing the process design, operating procedures, preventive measures, or precautionary instructions for process operators. An effective PSKM System could have also identified the subject matter experts and other technical specialists to conduct process design reviews, hazard analyses, and technical reviews.

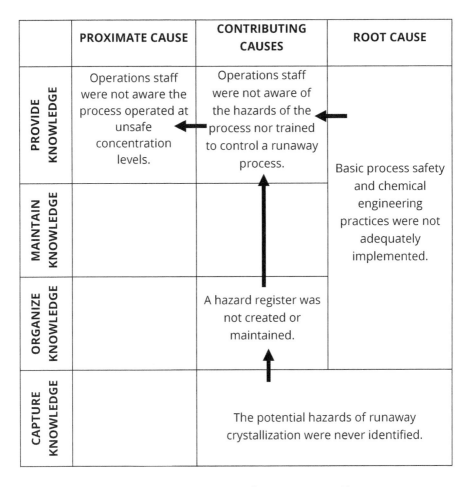

	PROXIMATE CAUSE	CONTRIBUTING CAUSES	ROOT CAUSE
PROVIDE KNOWLEDGE	Operations staff were not aware the process operated at unsafe concentration levels.	Operations staff were not aware of the hazards of the process nor trained to control a runaway process.	Basic process safety and chemical engineering practices were not adequately implemented.
MAINTAIN KNOWLEDGE			
ORGANIZE KNOWLEDGE		A hazard register was not created or maintained.	
CAPTURE KNOWLEDGE		The potential hazards of runaway crystallization were never identified.	

Figure 6-8 CSI Case Study PSKM Focus Chart

Figure 6-9 Aerial view of Concept Sciences Inc.

Source: [58]

6.2.3   Merrimack Valley, Massachusetts

"(...) locating accurate and up-to-date information about sensing lines was challenging because there was a shortage of information and confusion regarding what recordkeeping system would be used."

"(...) the successful execution of the South Union Street project was contingent upon employees remembering to transfer knowledge." [59]

1) Incident Summary:

In September 2018, a series of structure fires and explosions occurred after high-pressure natural gas was released into a low-pressure natural gas distribution system in the northeast region of the Merrimack Valley in the Commonwealth of Massachusetts. One person was fatally injured and 22 individuals, including three firefighters, were transported to local hospitals due to injuries; seven other firefighters incurred minor injuries. The fires and explosions damaged 131 structures, including at least 5 homes that were destroyed in the city of Lawrence and the towns of Andover and North Andover. Most of the damage occurred from fires ignited by natural gas-fueled appliances; several of the homes were destroyed by natural gas-fueled explosions.

The following excerpt is from the National Transportation Safety Board report [59]:

Natural gas distribution systems deliver natural gas to customers for heating, cooking, lighting, and other uses. A basic distribution system has three elements: (1) natural gas mains that transport natural gas underground, (2) service lines that deliver natural gas from the mains to customers, and (3) meters that measure the quantity of natural gas used by each customer. Customer piping takes natural gas from the meter to customer's appliances where it is used. To minimize service interruptions, normal maintenance and natural gas distribution system upgrades are typically performed with the system operating.

Both low-pressure and high-pressure natural gas distribution systems are used to supply natural gas to customers. In a low-pressure natural gas distribution system, the natural gas in the mains is essentially the same pressure as the pressure provided to the customer's piping and used by the appliances. Natural gas is typically supplied to the mains from a high-pressure source through a regulator station that reduces the pressure to that required by the customers.

The low-pressure natural gas distribution system in the Merrimack Valley was installed in the early 1900s with cast-iron mains. The system used 14 regulator stations to supply natural gas to the mains and control pressure. The regulator stations each contained two regulators in series, a worker regulator, and a monitor regulator each with a sensing line that feeds back the pressure in the main to the regulator forming a redundant closed-loop control system. The worker regulator is the primary regulator that maintains the natural gas pressure, and the monitor regulator provides a redundant backup to the worker regulator. Each of the regulator stations reduced the natural gas pressure from about 75 pounds per square inch gauge (psig) (517 kPa) to 12 inches of water column, about 0.5 psig (3.4 kPa), for distribution through the mains and delivery to customers. Since the regulator stations are the primary means of pressure control in the low-pressure systems, an overpressure condition in a natural gas distribution system could affect all customers served by the system. This is an inherent weakness of a low-pressure natural gas distribution system.

In a high-pressure natural gas distribution system, the natural gas pressure in the main is substantially higher than that required by the customer. A pressure regulator is installed at each meter to reduce the pressure. These regulators incorporate an overpressure protection device to prevent over-pressurization of the customer's piping and appliances should the regulator fail.

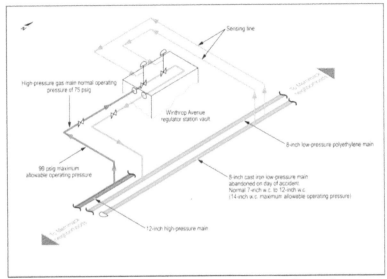

Figure 6-10 Merrimack Valley Gas Distribution Configuration

About 07:00 hours local time on the day of the incident, a Columbia Gas of Massachusetts' (CMA) construction coordinator, along with four employees of Feeney Brothers, arrived at Salem and South Union Streets in Lawrence, Massachusetts, to continue work on a CMA project to replace an existing cast-iron main with a polyethylene main. The crew completed the installation according to the CMA work plan, placed the new tie-ins into service, and isolated the existing cast-iron main shortly before 16:00 hours by closing valves on a 2-inch (5 cm) plastic bypass pipe between the cast-iron and polyethylene mains. Once the contractor crew isolated the cast-iron main, the natural gas pressure began to drop in the cast-iron main and in the regulator sensing lines. As the pressure dropped, the pressure regulators responded by opening further, increasing pressure in the natural gas distribution system. Since the Winthrop Avenue regulators no longer sensed system pressure, they fully opened, allowing high-pressure natural gas to be released into the low-pressure natural gas distribution system (Figure 6-10). The low-pressure sensing lines, shown in gold color in Figure 6-10, were not moved to the newly installed low-pressure polyethylene main from the out-of-service low-pressure cast-iron main. For this reason, there was no accurate indication of the actual pressure in the low-pressure polyethylene main when it was placed into service.

CMA uses three types of documents that are found in a work package and that are used to control the workflow of a construction project. Once these documents were complete, they were submitted to engineering management for approval. They include a Capital Design Job Order Checklist, a Capital Project Execution Workflow, and a Constructability Safety Review.

Constructability reviews are a recognized and generally accepted good engineering practice for the execution of professional design services and are intended to provide an independent and structured review of construction plans and specifications to ensure there are no conflicts, errors, or omissions (Kirby and others 1989). Two constructability reviews of the South Union Street project were signed on March 1, 2016, and January 6, 2017. The second constructability review was signed again on December 14, 2017. The constructability review form had a required signature line for the engineering and construction departments and an optional signature line for the Measurement and Regulation (M&R) department. The constructability review forms for the South Union Street project did not include signature(s) for representatives from the M&R department. Before the accident, the M&R department's participation in constructability reviews was on a case-by-case basis. For example, if the project involved changing the design or location of a regulator station or installing or replacing a regulator, M&R would likely be involved in the constructability review and meetings in the field.

Post-accident review of the engineering work package and construction documentation for the project identified some omissions. Although CMA used its project workflow process to develop, review, and approve the engineering plans, the work package did not consider the existence of regulator sensing lines connected to the distribution lines that were slated to be abandoned within the scope of work. This omission was not identified by any of the CMA constructability reviews including the one conducted in 2018 by NTSB [60]. In fact, none of the CMA workflow documents refer to natural gas distribution system pressure control nor do they refer to regulator control or sensing lines, and none of the documentation in the construction packages for the South Union Street project referred to sensing lines for regulator control. The 2018 constructability review document referenced pressure monitoring and stated that "if pressure rises/falls beyond these points, contact M&R."

2) PSKM Focus Chart:

The National Transportation Safety Board concluded the following:

- The probable cause of the over pressurization of the natural gas distribution system and the resulting fires and explosions was Columbia Gas of Massachusetts' (CMA) weak engineering management that did not adequately plan, review, sequence, and oversee the construction project that led to the abandonment of a cast-iron main without first relocating regulator sensing lines to the new polyethylene main.
- Contributing to the accident was a low-pressure natural gas distribution system designed and operated without adequate overpressure protection.
- There was no map available to emergency responders or information on how to shut the system down.

PSKM causes that led to the PSKM System failure in this incident consist of the following:

PSKM Proximate Cause:

The project work order package did not explicitly address sensing line locations or their relocation. There was no adequate plan for the abandonment of a cast-iron main without first relocating regulator sensing lines to the new polyethylene main.

PSKM Contributing Causes:

1. The Columbia Gas of Massachusetts constructability review process was not sufficiently robust to detect the omission of a work order to relocate the sensing lines.
2. It was difficult to find information on the configuration of the sensing lines.
3. The potential hazards of the sensing line configuration were not identified.
4. There was no hazard register for maintenance activities including line replacements for high-pressure to low-pressure gas lines.
5. The potential hazard of creating an interconnection that bypassed protections was not reviewed or identified.

6. Work crews replacing an existing cast-iron main with a polyethylene main were unaware of the high pressure in the supply line

PSKM Root Causes:

1. There was no Failure Mode and Effects Analysis (FMEA) conducted.
2. NiSource's (CMA's parent company) engineering risk management processes were deficient.
3. The Commonwealth of Massachusetts did not require a professional engineer's seal on public utility engineering drawings.
4. Although CMA crews were aware of issues that led to the incident, the knowledge was not transferred to new personnel assigned to the project. During the project, there was a nearly complete turnover of project personnel.

The PSKM Focus Chart in Figure 6-11 organizes all identified PSKM causes as gaps to capture, organize, maintain, and provide Process Safety Knowledge.

3) PSKM Benefits:

The incident at Merrimack Valley may have been less likely to occur with an effective PSKM system in place.

With an effective PSKM system in place, Columbia Gas of Massachusetts' engineering management may have adequately planned and reviewed the construction project and its potential hazards. A system to organize knowledge and a process for gas distribution operators to access that knowledge in the form of a hazard registry could have been developed. Hazard reviews to identify potential failures and propose recommendations to mitigate the failures could have been conducted. Information could have been provided to all low-pressure natural gas distribution system operators of the possibility of a failure of overpressure protection.

	PROXIMATE CAUSE	CONTRIBUTING CAUSES	ROOT CAUSE
PROVIDE KNOWLEDGE	The project work order package did not explicitly ← address sensing line locations or their relocation.	CMA's constructability review process was ← not sufficiently robust to detect the omission of a work order to relocate the sensing lines.	The Commonwealth of Massachusetts did not require a ← professional engineer's seal on public utility engineering drawings.
MAINTAIN KNOWLEDGE		It was difficult to find information on the configuration of the sensing lines.	
ORGANIZE KNOWLEDGE			
CAPTURE KNOWLEDGE		The potential hazard of creating an interconnection that ← bypassed protections was not reviewed or identified.	No risk assessment such as FMEA was conducted.

Figure 6-11 Merrimack Valley Case Study PSKM Focus Chart

Figure 6-12 Firefighters Battling Merrimack Valley Fire

Source: [61]

6.2.4   Tesoro Anacortes, Washington

"The refinery process safety culture required proof of danger rather than proof of effective safety implementation." [62]

1) Incident Summary:

On April 2, 2010, the Tesoro Refining and Marketing Company LLC ("Tesoro") petroleum refinery in Anacortes, Washington ("the Tesoro Anacortes Refinery"), experienced a catastrophic rupture of a heat exchanger in the Catalytic Reformer / Naphtha Hydrotreater unit ("the NHT unit"). The heat exchanger, known as E-6600E ("the E heat exchanger"), catastrophically ruptured because of High Temperature Hydrogen Attack (HTHA). Highly flammable hydrogen and naphtha at more than 500 degrees Fahrenheit ($°F$) (260 $°C$) was released from the ruptured heat exchanger and ignited, causing an explosion and an intense fire that burned for more than three hours. The rupture fatally injured seven Tesoro employees (one shift supervisor and six operators) who were working in the immediate vicinity of the heat exchanger at the time of the incident. At the time of publication, this was the largest fatal incident at a US petroleum refinery since the BP Texas City Refinery incident in March 2005.

The following excerpt [62] is from the U. S. Chemical Safety and Hazard Investigation Board (CSB) report:

> At the time of the release, the Tesoro workers were in the final stages of a startup activity to put the A/B/C bank of heat exchangers back in service following cleaning. The D/E/F heat exchangers remained in service during this operation. Because of the refinery's long history of frequent leaks and occasional fires during this startup activity, the CSB considers this work to be hazardous and non-routine. While the operations staff was performing the startup operations, the E heat exchanger in the middle of the operating D/E/F bank catastrophically ruptured.
>
> During normal operation at the Tesoro Anacortes Refinery, the A/B/C and D/E/F heat exchangers were all in use. Because of the original Shell Oil Company design and the process operating conditions, the heat exchangers would foul during operation; that is, they would develop a buildup of process contaminant byproducts both inside of the heat exchanger tubes, and

outside of the tubes. The fouling inhibited heat transfer between the tube-side and shell-side process fluid, thus reducing the heat transfer efficiency.

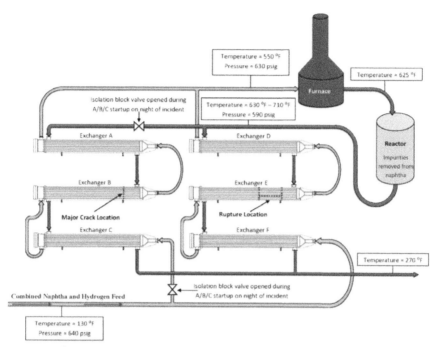

Figure 6-13 Tesoro Anacortes NHT Unit Heat Exchangers

Source: [62]

Because the heat exchangers fouled, they required periodic cleaning so that process temperature requirements could be maintained. Cleaning was typically required after about six months of continuous operation. When performing this cleaning, one bank of heat exchangers was taken out of service while the other bank continued operating. The cleaned heat exchangers would then be placed back into service by slowly introducing the hot naphtha and hydrogen feed into the heat exchangers. Because of a long history of frequent leaks and occasional fires when putting these heat exchangers back into service, startup, shutdown, and cleaning activities were a hazardous non-routine operation. By employing this non-routine operation, Shell Oil and Tesoro avoided a total shutdown of the NHT unit (Figure 6-13).

On March 28, 2010, five days before the incident, the A/B/C heat exchanger bank was taken offline so that the fouled tubes in each heat exchanger could be cleaned. The D/E/F heat exchanger bank and the rest of the NHT unit remained in operation. On March 31, 2010, the three-day maintenance cleaning activity was completed, and the equipment was reassembled and prepared for operation. On the evening of April 1, 2010, Tesoro initiated the startup of the A/B/C heat exchanger bank. The NHT unit was staffed in a typical manner, with one inside board operator who monitored the console and one outside operator.

The inside NHT operator and the outside NHT operator began the process of placing the heat exchangers back in service. The inside operator used a step-by-step task list for the startup process, physically checking off the steps on a hardcopy of the procedure while maintaining radio communication with the outside operator. Interviews conducted by the CSB indicate that the startup of the heat exchangers was a very difficult assignment for only a single outside operator. The startup procedure required manipulation of several isolation block valves, which necessitated a significant amount of manual effort to open.

These valves had to be gradually and concurrently opened, so the operator could not simply stay by each valve until it was fully opened or closed. Also, four steam lances were staged and ready for use during the start-up to mitigate any leaks or fires that might occur. These valves and steam lances were located at different positions in the vicinity of the A/B/C and D/E/F heat exchangers. At approximately 10:30 p.m., six additional Tesoro employees (five operators and one supervisor) joined the outside operator, at the request of the supervisor, to assist in bringing the A/B/C heat exchanger bank online. The startup procedure did not specify defined roles for these six additional personnel.

The operators continued the A/B/C heat exchanger bank startup as planned. Two leaks from the heat exchangers were reported during the startup. These leaks did not stop operations however, because leaks during startup of these heat exchangers were frequent and had become a "normal" part of the startup. Furthermore, based on past operating experience, these leaks were expected to cease when the heat exchangers reached typical operating temperature.

At 12:30 a.m. on April 2nd, while the seven outside personnel were still performing A/B/C heat exchanger bank startup operations, the E heat exchanger on the adjacent, in-service bank catastrophically ruptured. The pressure containing "shell" of the heat exchanger separated at weld seams, expelling a large volume of very hot hydrogen and naphtha. The naphtha and hydrogen likely auto ignited upon release into the atmosphere, creating a large fireball.

2) PSKM Focus Chart:

Summary of Incident Causes:

- Company policy did not prevent seven workers from being near the E Heat Exchanger [during maintenance nearby].
- The company was not formally aware of the poor mechanical integrity of the exchanger.
- The E Heat Exchanger was never inspected for potential HTHA damage.
- PHAs did not recognize hazardous conditions of the exchangers.
- Tesoro process safety culture was weak.
- Exchanger leaks during startup were normalized deviations.
- Exchangers were in service for 38 years.

PSKM causes that led to the PSKM System failure in this incident consist of the following:

PSKM Proximate Cause:

Severe High-Temperature Hydrogen Attack (HTHA) damage in the B and E heat exchangers was unknown to the operators, technical staff, and management.

PSKM Contributing Causes:

1. The E Heat Exchanger was never inspected for potential HTHA damage.
2. There was insufficient information to measure process conditions.
3. PHA teams did not identify fire scenarios even though the fires were commonly known.

PSKM Root Causes:

1. Adequacy of HTHA-preventing safeguards was never assessed.

2. Leaks were accepted as part of doing business, leading to a normalization of these deviations.

3. Tesoro's process safety culture was weak.

4. Tesoro did not know that the actual listed materials of construction for the B&E Heat Exchangers was not correct.

5. Tesoro relied on Nelson curves that API published but API did not have original data to confirm the curves and did not realize that the curves did not account for variables that affect HTHA, including Stress, Carbide Stability, Grain Size, Type of weld, and Time in Operation [79].

3) PSKM Benefits:

The incident at Anacortes Tesoro may have been less likely to occur with an effective PSKM system in place. The incident was caused by High Temperature Hydrogen Attack (HTHA) damage mechanism. The startup of the NHT heat exchangers was hazardous non-routine work yet leaks routinely developed that presented hazards to workers conducting the startup activities. PHA Teams repeatedly failed to identify the issue and respond with recommendations to ensure these hazards were controlled and the number of workers exposed to these hazards was minimized. Subsequent PHA revalidations did not address the leaks that occurred during startup and therefore included no recommendations to fix the leaks. PHA revalidations also failed to identify HTHA as a hazard for the shell of the B and E heat exchangers.

Had PSKM system been implemented, Tesoro could have captured the cause of the startup leaks, developed, and implemented an engineered solution to help prevent future incidents and shared the information with all personnel. Tesoro could also share information with API since API RP 941 contained a submittal sheet for companies to report their experience with HTHA [62]. Process knowledge could have been captured to eliminate the potential for fires that sometimes occurred during startup and minimized the number of workers in the field during startup needed to assist the one field operator perform the startup in a safe manner. Safeguards identified by PHA teams could have been implemented and the effectiveness of each safeguard could have been documented. Data could have been captured, provided, and maintained for technical experts' use during evaluation of the B and E heat exchangers for HTHA susceptibility. The mechanical integrity reviews periodically performed may have captured HTHA as a credible failure mechanism for the B and E heat exchangers. Source: [62]

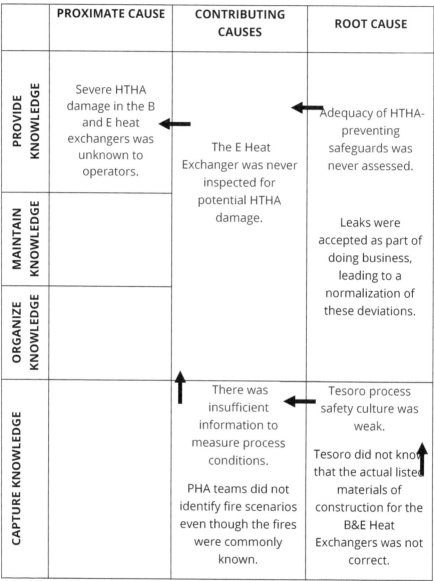

	PROXIMATE CAUSE	CONTRIBUTING CAUSES	ROOT CAUSE
PROVIDE KNOWLEDGE	Severe HTHA damage in the B and E heat exchangers was unknown to operators.	The E Heat Exchanger was never inspected for potential HTHA damage.	Adequacy of HTHA-preventing safeguards was never assessed.
MAINTAIN KNOWLEDGE			Leaks were accepted as part of doing business, leading to a normalization of these deviations.
ORGANIZE KNOWLEDGE			
CAPTURE KNOWLEDGE		There was insufficient information to measure process conditions. PHA teams did not identify fire scenarios even though the fires were commonly known.	Tesoro process safety culture was weak. Tesoro did not know that the actual listed materials of construction for the B&E Heat Exchangers was not correct.

Figure 6-14 Anacortes Tesoro Case Study Focus Chart

Figure 6-15 Image of the Fire After the Heat Exchanger Failure

6.2.5   Du Pont La Porte, Texas

"Due to the complex layout and emergency responders' lack of knowledge of the manufacturing building, the emergency responders had difficulty finding Operator 1." [63]

1) Incident Summary:

On November 15, 2014, approximately 24,000 pounds (11340 kg) of highly toxic methyl mercaptan was released from an insecticide production unit (Lannate® Unit) at the E. I. du Pont de Nemours and Company (DuPont) chemical manufacturing facility in La Porte, Texas. The release fatally injured three operators and a shift supervisor inside a manufacturing building. They died from a combination of asphyxia and acute exposure (by inhalation) to methyl mercaptan.

The following excerpt is from the U. S. Chemical Safety and Hazard Investigation Board report [63].

> The CSB determined that the cause of the highly toxic methyl mercaptan release was the flawed engineering design and the lack of adequate safeguards. Contributing to the severity of the incident were numerous safety management system deficiencies, including deficiencies in formal process safety culture assessments, auditing, and corrective actions, troubleshooting operations, management of change, safe work practices, shift communications, building ventilation design, toxic gas detection, and emergency response. Weaknesses in the DuPont La Porte safety management systems resulted from a culture at the facility that did not effectively support strong process safety performance. "The highly toxic methyl mercaptan release resulted from a long chain of process safety management system implementation failures stemming from ineffective implementation of the process safety management system at the DuPont La Porte facility."

> The CSB investigation viewed the chain of implementation failures as starting with the flawed engineering design of the $20 million nitrogen oxides (NOx) reduced scrubbed (NRS) incinerator, a capital project implemented in 2011. "DuPont La Porte had long-standing issues with vent piping to this incinerator because the design did not address liquid

accumulation in waste gas vent header vapor piping to the NRS, and DuPont La Porte did not fully resolve the liquid accumulation problem through hazard analyses or management of change reviews." Instead, to deal with these problems, daily instructions had been provided to operations personnel to drain liquid from these pipes to the atmosphere inside the Lannate® manufacturing building without specifically addressing the potential safety hazards this action could pose to the workers. DuPont La Porte's instructions did not specify additional breathing protection for this task.

"On the night of the incident, not realizing the hydrate blockage in the methyl mercaptan feed piping was cleared, workers went to drain liquid from the waste gas vent piping. They did not know that high pressure in the waste gas vent piping was related to the fact that liquid methyl mercaptan was flowing through the methyl mercaptan feed piping and into the waste gas vent piping." The chain further developed when the ineffective building ventilation system failed to be addressed after DuPont auditors identified it as a safety concern about five years before the incident. DuPont La Porte's management system did not resolve the process safety management recommendation (i.e., did not take corrective action) to address the building ventilation system. The ventilation design for the manufacturing building was based on flammability characteristics and did not take into consideration toxic chemical exposure hazards, even though the building contained two highly toxic materials: chlorine and methyl mercaptan. DuPont La Porte records indicated that the manufacturing building's dilution air ventilation design was based on providing sufficient ventilation to ensure that the concentration of flammable gases did not exceed 25 percent of the lower explosion limit (LEL). At the time of the incident, neither of the manufacturing building's two rooftop ventilation fans was working, despite an "urgent" work order written nearly a month earlier. Even had the fans worked, they probably would not have prevented a lethal atmosphere inside the building due to the large amount of toxic gas released.

DuPont La Porte's installation of a poorly designed, ineffective methyl mercaptan detection system inside the manufacturing building added another link to the chain. Neither the workers nor the public was protected by DuPont's toxic gas detection system on the night of the incident. The building where the workers died was not equipped with an adequate toxic gas detection system to alert personnel to the presence of hazardous chemicals.

First, DuPont La Porte set the detector alarms well above safe exposure limits for workers. Second, DuPont La Porte relied on verbal communication of alarms that automatically displayed on a continuously manned control board. Finally, DuPont La Porte did not provide visual lights or audible alarms for the manufacturing building to warn workers of highly toxic gas concentrations inside it. When a release caused a detector to register a concentration above the alarm limit, the toxic gas detection system did not warn workers in the field about the potential leak and the need to evacuate. Among other factors, this detection system contributed to workers' growing accustomed to smelling the methyl mercaptan odor in the unit. Additionally, when the toxic gas detectors triggered alarms, DuPont La Porte personnel investigated potential methyl mercaptan leaks without using respiratory protection. Personnel normalized unsafe methyl mercaptan detection practices by using odor to detect the gas, further deteriorating the importance or effectiveness of utilizing instrumentation in response to alarms signaling potential toxic gas releases.

Figure 6. Depiction of Path Taken by Operator 7 to Retrieve SCBAs for ERT When the Mini Pumper Truck Would Not Start. The hot zone was not controlled adequately to ensure that personnel did not enter a hazardous area. Source: Google Earth, with annotations by CSB.

Figure 6-16 Access Paths to SCBAs

Source: [63]

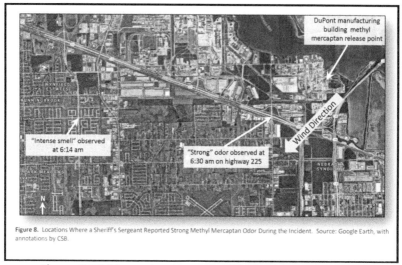

Figure 8. Locations Where a Sheriff's Sergeant Reported Strong Methyl Mercaptan Odor During the Incident. Source: Google Earth, with annotations by CSB.

Figure 6-17 Dispersion Impact of Methyl Mercaptan

Source: [63]

In the spring of 2014, the chain propagated when DuPont La Porte's interlock program did not require verification that interlocks that had been bypassed for turnaround maintenance were returned to service before the plant resumed operating. Because of this ineffective program, a bypassed interlock caused acetaldehyde oxime (AAO), a critical raw material, to become diluted, leading to a shutdown of the Lannate® Unit days before the November 2014 incident. During the shutdown, water entered the methyl mercaptan feed piping and, due to the cold weather, formed a hydrate (an ice-like material) that plugged the piping and prevented workers from restarting the unit. Furthermore, DuPont La Porte did not establish adequate safeguards after a 2011 DuPont La Porte process hazard analysis identified hydrate formation in this piping, revealing yet another link in the chain. When the hydrate formed, lacking safeguards to control the potential safety hazards associated with dissociating (breaking up) the hydrate (such as using heat tracing to prevent the hydrate from forming a solid inside the piping or developing a procedure to dissociate the hydrate safely), DuPont workers went into troubleshooting mode. Ineffective hazard management while troubleshooting the plugged methyl mercaptan feed piping formed yet another link in the chain and allowed liquid methyl mercaptan to flow into the waste gas vent header piping toward the NRS incinerator—a location where it was never intended to go.

DuPont La Porte did not fully resolve liquid accumulation in the waste gas vent header by the NRS incinerator. "Consequently, DuPont La Porte workers dealt with the common problem of liquid accumulation in the waste gas vent header on a routine basis by draining the liquid (line breaking) without an engineered solution or without ensuring the use of safety procedures or personal protective equipment." However, when the liquid drain valves were opened on November 15, 2014, flammable and highly toxic methyl mercaptan flowed onto the floor and filled the manufacturing building with toxic vapor. Once the methyl mercaptan release began, an ineffective emergency response program at La Porte contributed to the extent and duration of the chemical release, placed other workers in harm's way, and did not effectively evaluate whether the chemical release posed a safety threat to the public.

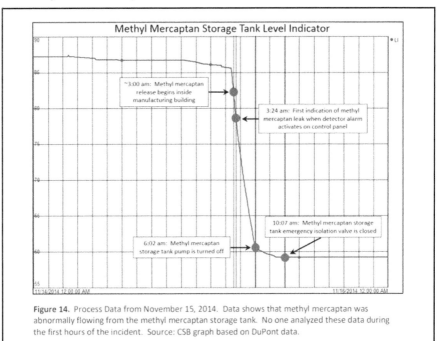

Figure 14. Process Data from November 15, 2014. Data shows that methyl mercaptan was abnormally flowing from the methyl mercaptan storage tank. No one analyzed these data during the first hours of the incident. Source: CSB graph based on DuPont data.

Figure 6-18 Loss of level in Methyl Mercaptan Storage Tank

Source: [63]

2) PSKM Focus Chart:

In its investigation of the November 15, 2014, DuPont La Porte incident, the CSB [63] concluded that:

- DuPont did not effectively respond to a toxic chemical release.
- DuPont's corporate process safety management system did not ensure that DuPont La Porte implemented and maintained an effective process safety management system.
- DuPont La Porte did not assess its culture for process safety in the site's Safety Perception Surveys or any other formal assessment program, allowing serious process safety deficiencies to exist at the site.

The CSB determined the cause of the highly toxic methyl mercaptan release was the flawed engineering design and the lack of adequate safeguards.

Summary of Incident Causes:

- The engineering design was flawed.
- There were insufficient safeguards to prevent a methyl mercaptan release or to notify operators of a release.
- DuPont did not address a liquid accumulation problem by standard PSM practices, but instead issued instructions to drain to atmosphere inside the building.

Figure 11. Photos of the Interior of the Lannate® Manufacturing Building. Source: CSB.

The CSB found that the Lannate® Unit's emergency planning and response manual did not have a building map or floor plan to aid emergency responders in understanding and navigating the manufacturing building during the incident. Emergency responders assigned Lannate® Unit operators—including the Board Operator—to draw maps of the manufacturing building to assist in the search. The Board Operator informed the CSB as follows:

> They got the big pad ... and had me sit down and start drawing pictures. And [the emergency responder] said, "We want it as detailed as you can get." So, I drew the building, I drew the stairwells, I drew the reactors, what reactors were where, you know, and coolers, you know, on each floor. That's what they wanted me to do, draw a description of each floor as well as I could ... so when they go in, they know where they're going. They can look, and of course three of [the missing operators were] found already, but they never did—at that time, they didn't know where [the fourth missing operator] was.... It took me about an hour and a half to draw stuff, you know.

Figure 6-19 Excerpt from CSB Investigation Report

Source: [63]

PSKM causes that led to the PSKM System failure in this incident consist of the following:

PSKM Proximate Causes:

1. Operators were not aware of the presence of methyl mercaptan in the liquids they were draining into the building.

2. Daily instructions had been provided to operations personnel to drain liquid from these pipes to the atmosphere inside the Lannate® manufacturing building without specifically addressing the potential safety hazards this action could pose to the workers.

PSKM Contributing Causes:

1. The building was not equipped with an adequate toxic gas detection system thus contributing to a lack of knowledge of the safety of the atmosphere.

2. Ventilations fans were not functioning or were not effective

3. DuPont La Porte did not fully resolve the liquid accumulation problem through hazard analyses or management of change reviews.

PSKM Root Causes:

1. Ventilation design did not identify toxic exposure as a major design consideration, thus leading to the potential for venting hazardous vapors into the building.

2. The design did not address liquid accumulation in waste gas vent header vapor piping to the NRS incinerator.

	PROXIMATE CAUSE	CONTRIBUTING CAUSES	ROOT CAUSE
PROVIDE KNOWLEDGE	Operators were not aware of the presence of methyl mercaptan in the liquids that they were draining into the building.	The building was not equipped with an adequate toxic gas detection system thus contributing to a lack of knowledge of the safety of the atmosphere.	Ventilation design did not identify toxic exposure as a major design consideration.
MAINTAIN KNOWLEDGE	Daily instructions had been provided to operations personnel to drain liquid from these pipes to the atmosphere inside the Lannate® manufacturing building without specifically addressing the potential safety hazards this action could pose to the workers.	DuPont La Porte did not fully resolve the liquid accumulation problem through hazard analyses or management of change reviews.	
ORGANIZE KNOWLEDGE			
CAPTURE KNOWLEDGE			Design did not address liquid accumulation in waste gas vent header vapor piping to the NRS incinerator.

Figure 6-20 DuPont LaPorte Case Study Focus Chart

3) PSKM Benefits:

The incident at DuPont may have been less likely to occur if there was an effective PSKM system in place. An implemented PSKM process could have captured the information regarding the liquid accumulation in the waste gas vent header piping to the incinerator and provided the information to subject matter experts to investigate, mitigate, and fully resolve either in PHAs or MOC reviews. Results could have been shared with all personnel. In addition, DuPont could have captured the breathing protection PPE requirement to prevent exposure to methyl mercaptan, shared the hazard information with all personnel, and trained all personnel in PPE use. In addition, the task to daily drain liquid from the waste gas vent piping to the atmosphere could have been eliminated to help prevent exposure. The captured information may have promoted a knowledge sharing culture.

6.2.6   Buncefield Oil Storage Depot

"At Buncefield the designers, manufacturers, installers and those involved in maintenance did not have an adequate knowledge of the environment in which the equipment was to be used." [64]

1) Incident Summary:

On the night of Saturday 10 December 2005, Tank 912 at the Hertfordshire Oil Storage Limited (HOSL) part of the Buncefield oil storage depot was filling with petrol. The tank had two forms of level control:

1. A gauge that enabled the employees to monitor the filling operation.

2. An independent high-level switch (IHLS) which was designed to automatically close transfer valves if the tank was overfilled.

The first gauge stuck and the IHLS was inoperable. There was therefore no means to alert the control room staff the tank was going to overfill. Eventually large quantities of petrol overflowed from the top of the tank. A vapor cloud formed which ignited causing a massive explosion and a fire that lasted five days.

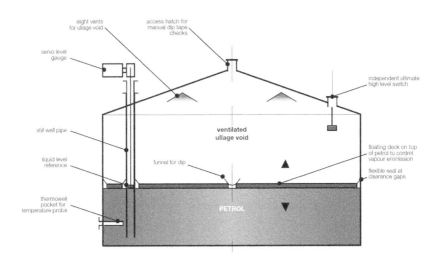

Figure 6-21 Buncefield Oil Storage Depot

Figure redrawn from The Buncefield Incident 11 December 2005: The final report of the Major Incident Investigation Board [81].

This incident was caused when an automatic fuel gauge in Tank 912 stopped functioning when the tank was being filled. The safety systems in place to shut off the supply to the tank to prevent overfilling failed to operate. The tank overfilled forming a vapor cloud resulting in a massive explosion and fire. Prior to the incident, the level gauge "flatlined" at least 14 times while the tank was either being filled or emptied. Service reports referenced adjustments made to the gear but the needed changes to the drum bearings were not completed. The gauge continued to stick requiring repeat maintenance. Root cause analysis was not performed. No consideration appears to have been given to replacing the level instrument. There was no evidence of an effective fault logging system or means to escalate a persistent problem to a more senior level. In 2004, the failed high-level switches were replaced and were treated as an "in kind" replacement. Opportunities were missed to ensure the changes necessary with the new switch had been properly reviewed and understood and documented. In all these areas, the ordered and fitted switch differed from that initially installed on the tank. There was no evidence that change control procedures had been applied to this modification. Further, there was no standardized procedure for filling a tank. Each operator had their own approach, and reliance was placed on high-level alarms to warn of the need to change over to a new tank.

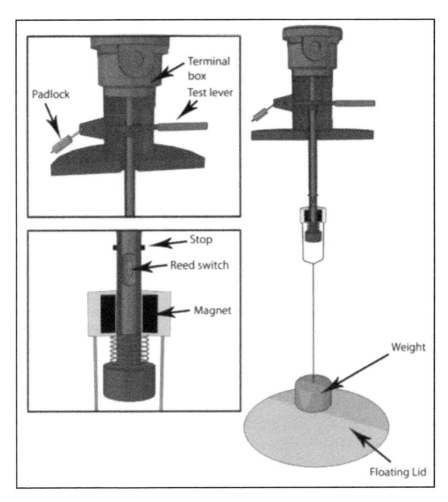

Figure 6-22 Description of the IHLS Design

IHLS – an Independent High-Level Switch. Source: [64]

2) PSKM Focus Chart:

Summary of Incident Causes:

- The level gauge had stuck intermittently after the tank was serviced in August 2005.
- The IHLS needed a padlock to retain its check lever in a working position that was not present.

- Management systems in place at Hertfordshire Oil Storage Ltd (HOSL) relating to tank filling were both deficient and not properly followed, even though the systems were independently audited.
- Pressures on staff had been increasing before the incident. The site was fed by three pipelines, two of which control room staff had little control over in terms of flow rates and timing of receipt. This meant that staff did not have sufficient information easily available to them to precisely manage the storage of incoming fuel.
- Throughput had increased at the site. This situation put more pressure on site management and its staff and further degraded their ability to monitor the receipt and storage of fuel. The pressure on staff was made worse by a lack of engineering support from the Head Office.
-

PSKM causes that led to the PSKM System failure in this incident consist of the following:

PSKM Proximate Causes:

> 1.   Operators did not have sufficient information easily available to them to precisely manage the receipt and storage of incoming fuel.
> 2.   There was inability to effectively monitor the receipt and storage of fuel.
> 3.   Maintenance was not aware that failing to secure the test switch with a padlock could cause an issue.

PSKM Contributing Causes:

1. There was no recognition the tank level flatlined below alarm conditions (level gauge stuck) and the tank continued to fill.
2. There was an Ineffective fault logging process and lack of maintenance regime to reliably respond to faults.
3. There was no effective action to find the root cause of the faulty gauge or to repair the fault completely.
4. Management of Change was bypassed with the replacement of the failed ILHS and thus the knowledge of changed functions was not captured or transmitted to operators.

PSKM Root Causes:

1. Increased throughput put more pressure on site management and its staff.
2. There was lack of engineering support from the Head Office.

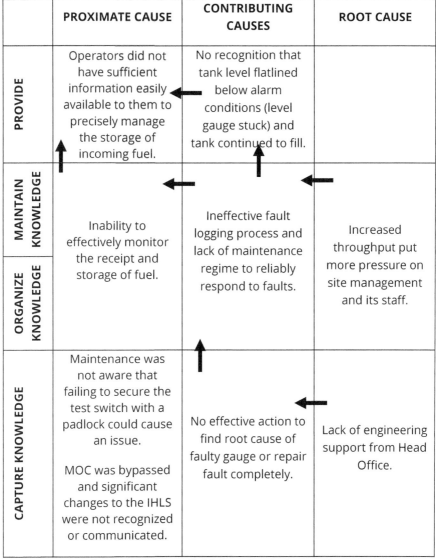

Figure 6-23 Buncefield Case Study Focus Chart

3) PSKM Benefits:

The incident at Buncefield may have been less likely to occur with an effective PSKM system in place, thus resulting in a safer workplace for the employees and a safer environment for the community.

Had a PSKM system been implemented, Buncefield could have captured the tank level instrument gauges including the independent high-level switch as safety critical instrumentation. The initial level gauge failure could have been identified as a high potential near miss and the information shared with operating and maintenance personnel. The near miss could have been recorded in their incident tracking system and been investigated to identify root causes and actions to prevent recurrence. If a repeat failure occurred, it could have been escalated to upper management and logged as a PSK KPI trending in the negative direction requiring immediate performance correction. Personnel assigned to maintain and repair critical equipment could have been able to capture and interpret data to understand and correct negative trends and to perform recommended procedures to avoid losses. The PSKM System would include a formal Management of Change assessment to identify, review, assess and approve all modifications to equipment. The information about the replaced high-level switches would require an update in PSI. In addition, Buncefield could have had a standard written procedure to provide operators with the correct instructions for filling a tank which could have eliminated operators from having their own approach to guessing when to stop filling.

As a result of the incident, extensive and detailed modeling was performed that provided knowledge not available at the date of the incident. The study identified that deflagration could lead to a detonation under specific circumstances. In the case of the Buncefield incident those circumstances included the presence of dense vegetation under piping racks that created congestion that contributed to the potential for a detonation [65]. If a PSKM SME Community of Practice existed this is the kind of information they could capture and communicate to others.

Figure 6-24 Aerial View of Buncefield Fire

Source: [66]

6.3 Key Factors from PSKM Success Stories in Other Industries

In the previous section, example failures of the PSKM system were identified and presented. In this section, key factors for successful implementation and/or operation of an effective PSKM System in other industries are provided. The chemical process industry can learn about successful implementation of knowledge management systems from success stories from other industries. Morrissey identified key factors in managing knowledge systems in the pharmaceutical and energy industries [67].

In the pharmaceutical industry, the identification of existing knowledge flow is key to designing an effective system. When implementing an effective system based on the knowledge flow, three different support functions are required: 1) communities of practice, 2) continuous improvement reviews, and 3) regular transfer of knowledge as case studies at the division level. These support functions require a comprehensive investment in document management, storage, and retrieval system. The document management system also needs to allow for locating a subject matter expert when needed. Finally, senior management support is crucial for implementing and sustaining such an extensive knowledge flow management system.

In the energy industry, the successful implementation of a knowledge management system requires various support functions such as communities of practice, significant training, and storytelling techniques. Again, the maintenance of PSKM in the energy industry depends on investing in a content management system.

Success factors have also been analyzed for knowledge management (KM) in the United States (US) Armed Forces [68]. Like the chemical process industries, the US Armed Forces need to manage a large volume of highly technical information while experiencing a high number of personnel changes to successfully achieve their mission. Key Practices in the Armed Forces include:

- Use of the Create, Craft, Choose, Promote and Organize (C-3PO) Framework which encompasses the tenets of successful change management, along with KM best practices and activities. The C-3PO Framework is deeply rooted in KM literature and coupled with thorough analysis of successful KM best practices.
  1. Create a KM Vision
  2. Craft a Strategy
  3. Choose KM Activities
  4. Promote Knowledge Sharing
  5. Organize KM Processes around Strategy
- Deployment and use of the Armed Forces Knowledge Management System, an information intranet tool, and an expertise locator

Further, KM control has been fully established in the U.S. Air Force, U.S. Army, and U.S. Navy. These branches of the military use four (4) compulsory elements consisting of "a predetermined set of targets, a means of measuring current activity, a means of comparing current activity with each target, and a means of correcting deviations from the targets". The four elements ensure implementation of KM strategy measures current performance and "guides the organization toward its changing image of the futures" in terms of KM [69].

Finally, some key factors include the following:

- Communities of Practice are fundamental to the success of Knowledge Management.
- Organizational, methodical, and procedural integration are key.
- Cultural integration is critical.

6.4   Understanding Knowledge Management System Failures

Common causes of knowledge management system failures can be presented using a Bow Tie diagram. The following is an example Bow Tie diagram and terms that apply to these types of diagrams [70].

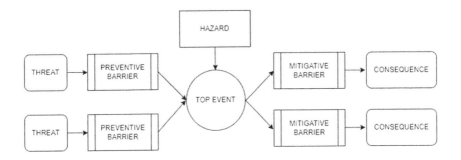

Figure 6-25 Example Bow Tie

Bow Tie Diagram - A risk diagram showing how various threats can lead to a loss of control of a hazard and allow this unsafe condition to develop into several undesired consequences. The diagram can also show all the barriers and degradation controls deployed.

The definitions for Hazard and Top Event are the approved definitions for use in this Guidelines book only.

Hazard – Any activity that includes, modifies, or manages the use of an inherent chemical or physical characteristic that has the potential for causing damage to people, property, or the environment.

Top Event - The loss event or other undesired event, otherwise known as the point of loss of control.

Threat - Any indication, circumstance, or event with the potential to cause the loss of, or damage to, an asset. Threat can also be defined as the intention and capability of an adversary to undertake actions that would be detrimental to critical assets.

Barrier - A control measure or grouping of control elements that on its own can prevent a threat developing into a top event (prevention barrier) or can mitigate the consequences of a top event once it has occurred (mitigation barrier). A barrier must be effective, independent, and auditable.

Preventive Barrier - A barrier designed to interrupt the chain of events leading up to a loss event, given that an initiating event has occurred.

Mitigative Barrier - A barrier designed to interrupt the chain of events after a loss event, given that there has been a loss of containment of a hazardous material or energy.

Consequence - The undesirable result of a loss event, usually measured in health and safety effects, environmental impacts, loss of property, and business interruption costs.

A Bow Tie diagram is read in the following manner:

The Hazard is an activity that is being assessed. The Top Event (is one of) the way(s) the activity can fail (this can be thought of as the point of loss of control). Threats are any events that can independently initiate the Top Event, and which will lead directly to the Top Event if there are no effective barriers to prevent it. The Consequence is an unwanted or negative outcome of a Top Event if no effective barriers are in place to prevent or mitigate it. Bow Tie diagrams are typically presented in graphic format and often contain multiple threats and multiple consequences with their attendant barriers. Note that multiple barriers are often present along any threat or consequence line. In the following examples, the barriers have been combined for clarity.

The following Bow Tie diagrams present commonly recognized failures in knowledge management and possible countermeasures (barriers) [71]. There are two main threat types: Inadequate System Design and Inadequate Attention to Human Nature. The consequences of failure of the PSKM method are Failure to Capture, Failure to Organize, Failure to Maintain, and Failure to Provide.

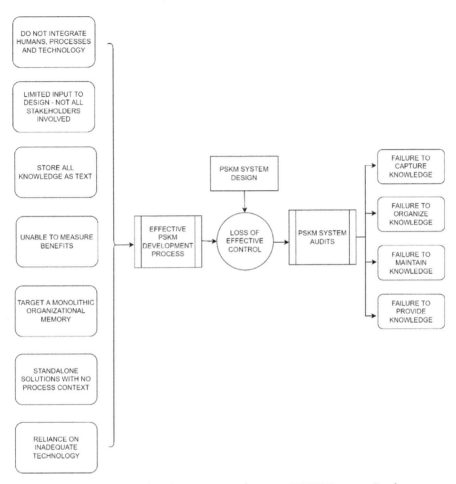

Figure 6-26 Bow Tie Diagram - Inadequate PSKM System Design

This Bow Tie shows the hazard represented by inadequate system design to a successful PSKM system. There are many potential threats to developing a PSKM System. As explained in Chapter 4, an effective PSKM development process plan coupled with strong leadership and management support can overcome these threats. If the PSKM System fails to deliver the desired outcomes, then PSKM System audits (See Chapter 5) can be used to identify the deficiencies and develop actions to mitigate or eliminate them.

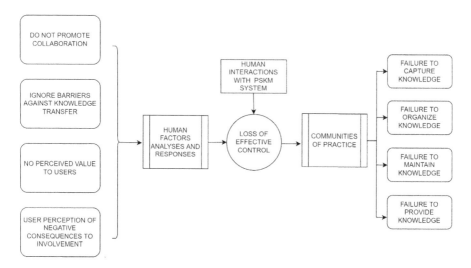

Figure 6-27 Bow Tie Diagram - Inadequate Attention to Human Factors

This Bow Tie shows the hazard represented by and inadequate approach to dealing with human factors affecting a successful PSKM system. There are many potential human factors threats to a PSKM System. As explained in Chapter 4, an effective PSKM management plan coupled with strong leadership and management support can overcome these threats. If the PSKM System fails to deliver the desired outcomes, then Communities of Practice (See Chapter 5) can operate to identify the deficiencies and develop actions to mitigate or eliminate them.

6.5   Chapter Summary

This chapter presented select case studies and lessons learned from chemical process safety incidents where it was evident from the investigations that PSKM was absent, poorly designed, poorly implemented, or poorly sustained within the organization. A new way to learn from incidents was introduced using the concept of the PSKM Focus Chart. The PSKM Focus Chart presents a methodology to look at an incident from a PSKM perspective and to identify correctable deficiencies in PSKM to prevent recurrence. The Case Study Summaries discussed in Section 6.2 demonstrate the importance of PSKM in process safety. The focus chart developed for each case study identifies failures to either Capture, Organize, Maintain or Provide Process Safety Knowledge to the right people at the right time. The lack of PSKM contributed to the significant incidents. These significant incidents shared common themes such as hazards not known or inadequate training. In Section 6.3, key factors in successfully managing knowledge systems were presented from the pharmaceutical, energy industries and the Department of Defense. The key factors that were identified included communities of practice and best practices sharing. In Section 6.4, common causes of Knowledge Management system failures were illustrated in Bow Tie diagrams.

6.6   Introduction to the Next Chapter

In Chapter 7, this Process Safety Knowledge Management (PSKM) book is summarized. The chapter presents observations of the observed failure over time of historical Process Safety Management programs to capture, organize, maintain, and provide critical safety management knowledge. Also presented are summaries of each preceding chapter and an ending restatement of the objectives of this book to share industry-leading best practices on PSKM and to provide a framework to help organizations to develop an effective PSKM program or to enhance their existing program.

7 Summary of Process Safety Knowledge Management (PSKM)

"The major problem with the chemical industry and indeed, with other industries, is the way accidents are investigated; reports are written, circulated, read, filed away and then forgotten. And then ten years later, even in the same company, the accident happens again. There is a saying that organizations have no memory; only people have memory and once they leave the plant, the accident that occurred there is forgotten about [72]."

Trevor Kletz (1922 – 2013)

Initially, process safety was based on word of mouth, such as limited passing of tidbits of experience handed down from one to another, almost like apprentices in trades. Before the 1990s, Process Safety Knowledge (PSK) existed in the organization as a core competency of Chemical or Process Engineers. Analysis of serious process safety events showed that while process safety information (PSI) resided within an organization, it did not consistently turn into knowledge at the operational level. The right knowledge did not get to the right people at the right time.

In Europe, two chemical plant accidents occurred in the 1970s that led to the adoption of major legislation aimed at preventing and controlling accidents: the Flixborough Accident (1974) and the Seveso Accident (1976). The major legislation known as the Seveso Directive was adopted in 1982. The goal of the Seveso Directive was to ensure the safety of workers, the surrounding communities and the environment's natural resources. The Seveso Directive placed an emphasis on prevention, preparedness and response to accidents involving hazardous substances at industries in the European Union.

The Seveso Directive was later amended to incorporate lessons learned from chemical incidents that occurred in the 1980s such Bhopal (1984), Chernobyl (1986), and Piper Alpha (1988).

These major incidents, including the explosion in Pasadena, Texas (1989), accelerated the development of the *Process Safety Management (PSM) of Highly Hazardous Chemicals* standard by the United States Occupational Safety and Health Administration (OSHA). The US OSHA standard was issued in 1992.

The Guidelines for Risk Based Process Safety (RBPS) [7] book published by AIChE CCPS in 2007 highlighted the need for PSK and its inclusion in a wider safety management system. The RBPS book links PSK to risk assessment and other risk management tools. This book advances the concepts introduced in the RBPS book to help organizations ensure knowledge and wisdom developed within people is not lost when people change their roles within or leave the organization.

PSKM is a system for capturing, organizing, maintaining, and providing the right Process Safety Knowledge to the right people at the right time to improve process safety in an organization.

Chapter 1 introduced the concept of PSKM and the Data-Information-Knowledge-Wisdom pyramid. Using the Plan-Do-Check-Act cycle within the context of PSKM facilitates movement from simply providing information to people who operate and maintain processes to transferring knowledge and developing wisdom in the very same people. The transition from data to wisdom can be represented by the DIKW pyramid (repeated in this Chapter as Figure 7-1 .

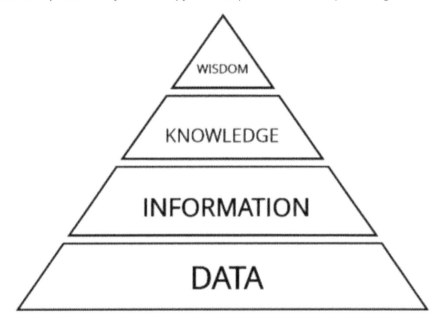

Figure 7-1 Data-Information-Knowledge-Wisdom (DIKW) pyramid

Organizations that operate at the bottom of this pyramid tend to be very reactive. As organizations move up the pyramid from a basic information-based approach to a more advanced, knowledge-based approach, they become more independent and less reactive. By making this transition, the organization has developed personal and/or group knowledge to operate more efficiently. The final move to organizational wisdom on the pyramid transforms the organizational culture to interdependence where understanding of process operations is deep and more effective.

Chapter 2 presented the Business Case for PSKM. There is a cost component associated with an effective PSKM System. An effective PSKM System includes three (3) phases: Generating, Retaining and Sharing Process Safety Knowledge. Financial benefits are realized when businesses run efficiently. By adopting PSKM, efficiencies are realized by generating, retaining, and sharing lessons learned from lost-time incidents and business interruption throughout the organization, so as not to repeat them. Additional business benefits resulting from the adoption of a PSKM strategy include reduced costs associated with process hazard analysis studies, a good knowledge-sharing system to minimize the chance for mistakes, and an effective knowledge management system that uses lessons learned.

Chapter 3 tied together PSKM and Risk Based Process Safety (RBPS). There are six elements of the PSKM cycle that impact the four pillars of RBPS. The PSKM System funnels captured knowledge and information through the four pillars of RBPS. As the captured knowledge is organized, it becomes refined and readily available to specific people when it is needed. While all RBPS elements interact with PSKM, four of the elements are especially important: Process Knowledge Management, Operating Procedures, Training, and Management of Change. Knowledge is generated in Process Knowledge Management. Knowledge is shared in Operating Procedures and Training. Finally, knowledge remains evergreen through Management of Change.

Chapter 4 provided suggestions for developing and implementing a Process Safety Knowledge Management strategy in any organization. This chapter provides a detailed discussion of the PSKM system implementation, including key steps and examples in developing a PSKM program. We also provided potential obstacles to overcome during implementation. A successful PSKM System is comprised of four elements: Capture, Organize, Maintain and Provide. A process for capturing and organizing the knowledge is presented together with potential barriers to PSKM along with strategies to drive the PSKM and establish a PSKM culture. Tools for data management and benefits an organization realizes using PSKM tools are discussed. Resources are required to capture, organize, maintain, and provide the PSK for the whole organization. Having clearly defined roles and responsibilities enables management to identify the types of qualified people needed to successfully implement the PSKM System. An example PSKM RACI Chart is provided.

Chapter 5 discussed steps for maintaining and improving Process Safety Knowledge Management in any organization, including ten (10) tools that can be used to maintain PSK, as well as an example logic model illustrating the relationship between activities and goals. KPIs for measuring PSKM performance are identified and explained. Core components for long-term success of the PSKM program are described. A process is presented for assessing an organization's PSKM maturity and conducting a PSKM gap assessment and audit. For organizations that use, utilize, or make use of software to maintain PSK, a checklist is provided to evaluate the essential requirements needed. Finally, data management technologies have been discussed for effective communication, collaboration, and information storage.

Chapter 6 introduced the PSKM Focus Chart approach to analyze a few select case studies and lessons learned from chemical process safety incidents where it was evident from the investigations that PSKM was absent, poorly designed, poorly implemented, or poorly sustained within the organization. Companies understand there is much to be learned from analyzing and reviewing their own incidents. When the organization is also capable of learning from others' past experiences, then true organizational learning can be achieved that directly and positively impacts the PSKM System. As the process safety pioneer, Jesse C. Ducommun (former Vice President of American Petroleum Institute in 1964) once said: "It should not be necessary for each generation to rediscover the principles of process safety which the generation before discovered. We must learn from the experience of others rather than learn the hard way. We must pass on to the next generation a record of what we have learned." [73]

This book is intended to be a resource for sharing industry-leading best practices on PSKM and for providing a blueprint to help organizations develop an effective PSKM program or enhance their existing program. In addition, the works cited are provided to assist all organizations develop and implement a structured approach to Process Safety Knowledge Management that can ultimately add business value and improve safety and lead to creation of a robust, sustainable PSKM culture.

References

[1]   CCPS, *CCPS Process Safety Glossary,* aiche.org/ccps, New York, NY USA: Center for
        Chemical Process Safety, 2023.

[2]   J. S. Carroll, "Knowledge Management in High-Hazard Industries Accident
        Precursors as Practice," in *Accident Precursor Analysis and Management: Reducing
        Technological Risk Through Diligence*, The National Academies Press, 2004.

[3]   Sun Associates, *Creating Project Logic Maps (Using the Logic Model),* sun-
        associates.com/logicmodels, North Chelmsford, MA: Sun Associates, 2018.

[4]   J. G. John Girard, "Defining knowledge management: Toward an applied
        compendium," *Online Journal of Applied Knowledge Management,* vol. 3, no. 1, pp.
        1 -20, 2015.

[5]   C. O'Dell, C. J. Grayson and w. N. Essaides, If Only We Knew What We Know; The
        Transfer of Internal Knowledge and Best Practice., The Free Press, 1998.

[6]   CCPS, Guidelines for Process Safety Documentation, Hoboken, NJ USA: John
        Wiley & Sons, 1995.

[7]   CCPS, Guidelines for Risk Based Process Safety (RBPS), New York, NY: AIChE,
        Wiley & Sons, 2007.

[8]   J. Rowley, "The wisdom hierarchy: representations of the DIKW heirarchy.,"
        Journal of Information and Communication Science, vol. 2, no. 33, pp. 163 - 180,
        2007.

[9]   T. H. Davenport and P. Laurence , Working Knowledge: How Organizations
        Manage What They Know, Massachusetts: Harvard Business School Press, 2000.

[10]   D. Podgórski, "The Use of Tacit Knowledge in Occupational Safety and Health
        Management Systems," *International Journal of Occupational Safety and
        Ergonomics (JOSE),* vol. 16, no. 3, p. 283 – 310, 2010.

[11] B. C. Anderson, "Developing Organizational and Managerial Wisdom – 2nd Edition," Kwantlen Polytechnic University, kpu.pressbooks.pub/developingwisdom/, Canada, 2020.

[12] CCPS, The Business Case for Process Safety, 2018.

[13] W. G. Bridges, A. I. Dowell, J. Thomas and P. Casarez, "Common Hurdles, Benefits, and Costs for Fully Implementing Process Safety Worldwide," in *Global Congress on Process Safety*, 2017.

[14] P. Leavitt, *Applying Knowledge Management to Oil and Gas Industry Challenges*, American Productivity & Quality Center, 2002.

[15] *The Business Journals*, New York Times, March 19, 2020.

[16] R. S. (Editor-In-Chief), *Daily COVID-19 Updates: March 24*, industryweek.com/operations, Independence, Ohio: Industry Week, March 24, 2020.

[17] J. Chosnek, "The Taxonomy (Classification) of Process Safety," in *Mary Kay O'Connor Process Safety Center International Symposium*, College Station, 2010.

[18] U.S. Chemical Safety and Hazard Investigation Board, "T2 Laboratories, Inc. Runaway Reaction, Report 2008-3-I-FL," US Chemical Safety Board, Washington, D.C. USA, 2009.

[19] J. Chosnek, "Lessons learned - How to make them stick," *Process Safety Progress*, vol. 40, no. 2, p. e12198, 2021.

[20] B.P., "Final Accident Investigation Report, Isomerization Unit Explosion Final Report," 2005.

[21] L. Edvinsson, Corporate Longitude: What You Need to Know to Navigate the Knowledge Economy, Financial Times Management, 2002.

[22] M. Gelfand, Rule Makers, Rule Breakers: Tight and Loose Cultures and the Secret Signals that Direct our Lives, New York: Simon & Schuster, 2018.

[23]　CCPS, "Risk Based Process Safety Overview," AICHE, New York City, 2014.

[24]　CCPS, *CCPS Process Safety Beacon,* aiche.org/ccps: Center for Chemical Process Safety, 2023.

[25]　CSB, *Investigations,* csb.gov/investigations: U.S. Chemical Safety and Hazard Investigation Board, 2023.

[26]　EPSC, *EPSC Learning Sheets,* epsc.be: European Process Safety Centre, 2023.

[27]　PMI, "A Guide to the Project Management Body of Knowledge (PMBOK® Guide), 7th Edition," Project Management Institute, pmi.org, 2021.

[28]　L. Halawi, R. McCarthy and J. Aronson, "Success Stories in Knowledge Management Systems," *Issues in Information Systems,* vol. 18, no. 1, pp. 64-77, 2017.

[29]　R. S. Wurman, L. Leifer, D. Sume and K. Whitehouse, Information Anxiety 2, 2nd Edition, Hoboken, NJ: Que Publishing, Pearson, 2000.

[30]　W. Craig, "The Nature of Leadership in a Flat Organization," *Forbes,* no. October 23, 23 October 2018.

[31]　ASQ, "Best of Back To Basics: The Benefits of PDCA," *Quality Progress,* vol. 49, no. 1, p. 45, January 2016.

[32]　L. Ruzic-Dimitrijevic, "Risk assessment of knowledge management system," *Online Journal of Applied Knowledge Management,* vol. 3, no. 2, p. 12, 2014.

[33]　C. Hubert and B. Lopez, "Breaking the Barriers to Knowledge Sharing," APQC (American Productivity & Quality Center), www.apqc.org, 2013.

[34]　CCPS, "Guidelines for Hazard Evaluation Procedures, Third Edition," Wiley, 2008, p. 281.

[35]　J. Rasmussen, Information Processing and Human-machine Interaction: an approach to cognitive engineering, New York: Elsevier Science Ltd., 2016.

[36]   J. Chosnek, "Lifecycle Management of Protection Layers and Safeguards," in *2019 Spring Meeting and 15th Global Congress on Process Safety*, New Orleans, 2019.

[37]   CCPS, *Chemical Reactivity Worksheet (CRW)*, www.aiche.org/ccps: AIChE/CCPS, 2023.

[38]   M. H. Jackson and J. Williamson, "Knowledge Systems: Static Systems and Dynamic Processes," in *Communication and Organizational Knowledge: Contemporary Issues for Theory and Practice*, New York City, Routledge/Taylor & Francis, 2011, pp. 53-64.

[39]   V. D. Manriquez, "The Importance of Integrating Knowledge Management with Maintenance," *Mahinery Lubrication,* October 2014.

[40]   Society for Human Resource Management (SHRM), *Managing for Employee Retention,* SHRM, 2016.

[41]   J. Clear, *Mental Models: How to Train Your Brain to Think in New Ways,* jamesclear.com/feynman-mental-models.

[42]   CCPS, Guidelines for Hazard Evaluation Procedures, 3rd Edition, Hoboken, NJ USA: John Wiley and Sons, 2008.

[43]   CCPS, "Guidelines for Mechanical Integrity Systems," Wiley, 2006, pp. 5, 138-140.

[44]   C. Jackson, *Engineering Knowledge Management,* lifecycleinsights.com/tech-guide/engineering-knowledge-management: Lifecycle Insights, 2015.

[45]   CCPS, *CCPS Golden Rules of Process Safety,* www.aiche.org/ccps/publications/golden-rules-process-safety: AICHE, 2021.

[46]   L. C. Emmett, *Knowledge management systems: Maintenance on the front lines,* plantengineering.com/articles/knowledge-management- systems-maintenance-on-the-front-lines: Plant Engineering, 1991.

[47]   S. Wan, R. Evans, J. Gao and D. Li, "Knowledge Management for Maintenance, Repair and Service of Manufacturing System," in *International Conference on Manufacturing Research*, Southampton Solent University, UK, 2014.

[48]   Du Plessis and Boshoff, "Preferred communication methods and technologies for organizational knowledge sharing and decision making," *SA Journal of Information Management,* vol. 10, pp. 1-18, June 2008.

[49]   The Columbia Electronic Encyclopedia, 6th Edition, *Information storage and retrieval,* infoplease.com/encyclopedia: Columbia University, 2023.

[50]   A. Malloy, *Information System: Uses, Examples And Characteristics,* crgsoft.com/information-system-uses-examples-and-characteristics/: Collaborative Research Group, 2022.

[51]   G. Shah, "5 Steps to Run Accident Investigations Right," *Chemical Processing,* 30 September 2016.

[52]   J. Chosnek, "Organizing Knowledge for Improved Process Safety," in *Mary Kay O'Connor Process Safety Center International Symposium*, College Station, 2008.

[53]   T. Kletz, Lessons from Disaster: How Organizations Have No Memory and Accidents Recur, Gulf Professional Publishing, 1993.

[54]   U.S. Chemical Safety and Hazard Investigation Board, "Investigation Report Refinery Explosion and Fire at BP Texas City, Texas Report No. 2005-04-I-TX," US CSB, csb.gov, 2007.

[55]   National Institute for Occupational Safety and Health, *Work and Fatigue,* cdc.gov/niosh: National Institute of Occupational Safety and Health (NIOSH), 2023.

[56]   U.S. Chemical Safety and Hazard Investigation Board, "The Explosion at Concept Sciences: Hazards of Hydroxylamine, Case Study No. 1999-13-C-PA," US CSB, csb.gov, 2002.

[57]   Federal Emergency Management Agency, "Concept Sciences, Incorporated Hanover Township, Pennsylvania USFA-TR-127," 1999.

[58] D. P. Sheenan and A. Salamone, "20 years ago, a thunderous explosion shook the Lehigh Valley," *The Morning Call,* 20 February 2019.

[59] National Transportation Safety Board, "Pipeline Accident Report NTSB/PAR-19/02," 2019.

[60] National Transportation Safety Board, "Natural Gas Distribution System Project Development and Review (Urgent) PSR-18/02," NTSB, Washington, DC, 2018.

[61] B. Bergstein, *It was a suburban disaster,* bu.edu/articles/2019/merrimack-valley-gas-explosions-joe-albanese/: Bostonia, Boston University's Alumni Magazine, 2019.

[62] U.S. Chemical Safety and Hazard Investigation Board, "Catastrophic Rupture of Heat Exchanger, Report 2010-08-I-WA," 2014.

[63] U.S. Chemical Safety and Hazard Investigation Board, "Toxic Chemical Release at the DuPont La Porte Chemical Facility, No. 2015-01-I-TX," 2015.

[64] Major Incident Investigation Board (MIIB), "The final report of the Major Incident Investigation Board, Volumes 1 & 2," HSE, Bootle, 11 December 2008.

[65] Steel Construction Institute, "Buncefield Explosion Mechanism Phase 1 (RR718)," Health and Safety Executive, 2009.

[66] Buncefield Major Incident Investigation Board, "The Buncefield Incident 11 December 2005, Volume 2," UK HSE, Crown, London , 2008.

[67] S. Morissey, "The Design and Implementation of Effective Knowledge Management Systems," Ford Motor Company MBA Fellowship, 2005.

[68] P. R. Johnson, "Thesis: Developing A Knowledge Management Framework to Assist with Current USMC Information Management Practices," Naval Postgraduate School Monterey, California, September 2010.

[69] C. Minonne and G. Turner, Strategic Knowledge Measurement: An Integrative Approach: A Structured Guideline For Assessing Knowledge Management Performance, Suedwestdeutscher Verlag fuer Hochschulschriften, 2010.

[70]　CCPS and the Energy Institute, Bow Ties in Risk Management: A Concept Book for Process Safety, Hoboken, NJ USA: John Wiley & Sons, 2018.

[71]　R. O. Weber, "Addressing Failure Factors in Knowledge Management," *Electronic Journal of Knowledge Management,* vol. 5, no. 3, pp. 333-346, 2007.

[72]　U.S. Chemical Safety and Hazard Investigation Board, *Remembering Trevor Kletz,* youtube.com/watch?v=XQn5fL62KL8: csb.gov, 2013.

[73]　F. Gil and J. Atherton, "Can we still use learnings from past major incidents in non-process industries?," in *IChemE Symposium Series NO. 154*, 2008.

Index